THE HISTORY OF SCIENCE
FROM 1895 TO 1945

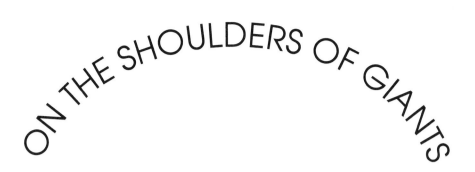

THE HISTORY OF SCIENCE
FROM 1895 TO 1945

Ray Spangenburg and Diane K. Moser

Facts On File, Inc.

On the cover: Marie Curie in her laboratory c. 1905 (The Bettmann Archive)

The History of Science from 1895 to 1945

Copyright © 1994 by Diane K. Moser and Ray Spangenburg

Facts On File, Inc.
11 Penn Plaza
New York NY 10001

Library of Congress Cataloging-in-Publication Data

Spangenburg, Ray, 1939–
    The history of science from 1895 to 1945 / Ray Spangenburg and Diane K. Moser.
      p. cm. — (On the shoulders of giants)
    Includes bibliographical references and index.
    ISBN 0-8160-2742-0
    1. Science—History—Juvenile literature. 2. Life sciences—History—Juvenile literature. I. Moser, Diane, 1944– .
II. Title. III. Series: Spangenburg, Ray, 1939– On the shoulders of giants.
Q163.S64   1994
509'.04dc20   93-26820

Text design by Ron Monteleone
Cover design by Semadar Megged
Composition by Facts On File, Inc./Robert Yaffe
Manufactured by the Maple-Vail Book Manufacturing Group
Printed in the United States of America

10 9 8 7 6 5 4

This book is printed on acid-free paper.

C O N T E N T S

*To the memory of
Albert Einstein and Niels Bohr,
great and gentle minds
who led the way into the 20th century*

ACKNOWLEDGMENTS

*B*ooks are a team effort, and this one is no exception. So many people have been generous with their time, talents and expertise in helping us with this book—both as we wrote it and in the past when it was only a dream—more than we can possibly name. We'd like to thank everyone for their help and encouragement, and a special thank you to: Gregg Proctor and the rest of the staff at the branches of the Sacramento Public Library for their tireless help in locating research materials. Shawn Carlson, of Lawrence Berkeley Laboratories, as well as Beth Etgen, educational director at the Sacramento Science Center, and her staff, for kindly reading the manuscript and making many helpful suggestions. Andrew Fraknoi of the Astronomical Society of the Pacific, for his help with research contacts and illustrations. Also for their help with illustrations: Jan Lazarus of the National Library of Medicine, Mark L. Darby of the Institute for Advanced Study, R. W. Errickson of Parke-Davis and Doug Egan of the Emilio Segrè Visual Archives. Thanks above all to Facts On File's editorial team, especially Nicole Bowen for her intelligent, energetic and up-beat management of the project; Janet McDonald for her eagle-eyed passion for detail; and James Warren for his vision and encouragement in getting it started. And to many others, including Jeanne Sheldon-Parsons, Laurie Wise, Chris McKay of NASA Ames, Robert Sheaffer and Bob Steiner, for many long conversations about science, its history and its purpose.

INTRODUCTION

AN OVERVIEW: 1895–1945

Throughout much of Europe, the British Isles and North America, the 20th century dawned as an era of great hope and promise. A bright future filled with prosperity seemed to stretch out endlessly ahead. And the hundreds of discoveries and inventions made during the preceding century—the fruits of science and technology—seemed to open up a vast treasure of possibilities in these last years of the 1800s and the first few years of the 1900s.

By 1880, Thomas Edison was illuminating the city of Menlo Park, New Jersey with sparkling electric lights. The Impressionist painters had held their first show in Paris in 1874. Progressive social welfare laws were established in Germany in 1885. The wonders of air travel seemed just around a corner—and by 1900 the Zeppelin dirigible would make its first voyage. A new private mode of transportation called the automobile promised to change everyone's everyday lives—especially in the vast expanses of the North American prairie, where getting to town or across the state or province became, like magic, far less time-consuming and cumbersome. Most people believed that this sort of magic was sure to continue forever.

For many people of the middle and upper classes the early years of the 20th century were good ones, years of peace, prosperity, and pleasant living. For the working class, industrialization and urbanization created problems, but these new social trends also promised opportunity and possibility. Expanding horizons for individuals came into view at every turn.

More people had the right to vote and influence their governments than ever before in history. In Britain, male members of the middle class were voting by 1832, working men by 1867. In the United States, by the 1890s several states had established the right to vote for women (although not at the federal level until 1920).

In England, Europe, Canada and the United States, especially, more people had an opportunity for education than ever before; and with education, it became possible for individuals of varied backgrounds to earn higher salaries, gain prestige and achieve a higher standard of living. In the United States, by 1900, 30 states required children to attend school, and the 800 high schools in existence in 1880 had expanded dramatically to 5,500 only 20 years later, with an increase in college attendance as well.

Not everything was quite as rosy as it seemed, however. And, as became apparent all too soon in the 20th century, many threats to civilization lay just beneath the waters most people perceived to be smooth. The world's economic condition was increasingly shaky and the nationalism that had seemed so healthy in the early 1800s—a stimulator of individual rights, independence and freedom—turned into a hungry monster, with national power as its overriding goal. Concern for individual freedoms became subordinated to national power and prestige, and war became the false proving ground of heroism and greatness. Some political thinkers, freely interpreting Charles Darwin's principle of "survival of the fittest," tried to apply Darwin's theory to human society and came up with Social Darwinism, which they used to justify everything from capitalism to communism, and many -isms in between, including racism, anti-Semitism and fascism.

Extreme nationalism incited powerful political rivalries and stepped-up military forces, and in 1914 the world exploded into war. What began as a result of long-held hostilities within Germany eventually drew most of the rest of the world into a bloody conflict. "The lamps are going out all over Europe," announced Edward Grey, the British foreign secretary, at the start of the war. "We shall not see them lit again in our lifetime." It was called the Great War, "the war to end all wars," and its cost was high. Many lives were lost, homes and industries destroyed. In one battle alone near Ypres, Belgium, more than 250,000 met their death. By war's end in 1918 more than 11 million people had died.

Meanwhile, empires toppled. In Russia, rule by the czars came abruptly to an end in 1917, to be replaced by a new, dictatorial, communist regime. And Russia's new communist government provoked an atmosphere of unease and fear throughout the world with its stated intent to overthrow capitalism and incite worldwide revolution. Throughout the rest of the world, many people who truly were oppressed or felt themselves to be oppressed by the social and economic conditions in their nations joined in sympathy with the communist cause, and the unrest became ever more widespread.

Deepening fear, joined with unhealed animosities in the wake of the Great War, led to strengthened extremist powers in Europe. In Asia, too, nationalism flared. By the 1930s, a second world war loomed inevitably on the path ahead.

Like the Great War, now known as World War I, the Second World War began in Europe, in Germany, where resentment, fascism and anti-Semitism had continued to grow after the defeat of 1918. Swept in by internal political discontent, fear of communism and the worldwide economic depression, an energetic, charismatic and arguably insane man named Adolf Hitler came to power, melding his personal thirst for glory with Germany's hunger for land, economic growth and prestige. Between the years of 1936 and 1945, 6 million Jews were murdered. Millions were thrown into concentration camps or fled, bereft of all their property and possessions, which had been confiscated by the German government. During the early years of Germany's aggression, most of the world's nations held back from war, still scarred by the brutal madness of the previous encounter. Finally, in 1939, Germany invaded Poland, and war broke out. By the end of it, Germany had joined forces with Italy to the south and Japan in the Pacific, while the Allies, led by the British Commonwealth, the United States, the Soviet Union and China, pitted themselves against the aggressors. By the end of the war more than 30 million people had died. It was the biggest, ghastliest and most devastating war the world had ever seen, and it left vast changes in its wake.

In science, at the turn of the century, a veritable burst of discoveries ushered in the modern scientific era. Between 1859 and 1895, Darwin's *Origin of Species* saw publication, Dmitry Mendeleyev organized the elements into a periodic table that showed their relationships and hinted at the existence of an atomic structure and Joseph Lister performed the first antiseptic surgery. From 1895 to 1912, a series of extraordinary scientific finds and theoretical breakthroughs would suddenly turn the world of physics topsy-turvy and thrust a new, modern scientific era upon the world. At the same time, these discoveries would place more power, both constructive and lethal, than ever before in the hands of human beings. The result in 1945—the concluding year of World War II and of this book—was a mushroom-shaped cloud that transformed forever the world in which we live and defined once again the complex and uneasy role of science in human history.

Our story begins with an enormously famous set of names: Wilhelm Röntgen, Marie and Pierre Curie, J. J. Thomson, Albert Einstein, Max

Planck, Ernest Rutherford, Niels Bohr. Theirs was a time of great vitality, excitement and confusion in the field of physics, and their work ranks among the most exciting in the history of science.

This book, like the four others in the series called On the Shoulders of Giants, looks at how people have developed a system for finding out how the world works, making use of both success and failure. We will look at the theories they put forth, sometimes right and sometimes wrong. And we will look at how we have learned to test, accept and build upon those theories, or to correct, expand or simplify them.

We'll also see how scientists have learned from others' mistakes, sometimes having to discard theories that once seemed logical but later proved to be incorrect, misleading or unfruitful. In all these ways they have built upon the shoulders of other men and women of science—the giants who went before them.

PART ONE

THE PHYSICAL SCIENCES

C H A P T E R   1

THE NEW ATOM: FROM X RAYS TO THE NUCLEUS

Wilhelm Conrad Röntgen [RUNT gen] was in for a surprise the evening of November 8, 1895. Working in his half-darkened

Wilhelm Conrad Röntgen, discoverer of X rays (National Library of Medicine)

laboratory at the University of Würzburg in Bavaria, he was suddenly distracted by a mysterious glow in one far-off corner of the room. He took a closer look. The strange gleam came from a piece of paper coated with barium platinocyanide, a substance, he knew, that glowed with an eerie luminescence when exposed to cathode rays. But this time, there could be no rays to reflect: The cathode ray tube he had been working with was covered with a heavy piece of cardboard, and, anyway it was clear across the room! Yet when he turned the cathode ray tube off, the paper stopped glowing. When he turned the tube back on, the light shone eerily again. He put his hand between the cathode ray tube and the coated paper. His hand cast a shadow in the light, and he could see the *bones in his hand!* He took the coated paper with him to another room, shut the door and pulled the blinds. It still glowed when the cathode ray tube was turned on. It stopped glowing when the tube was turned off. The mysterious rays that were causing the glow had actually passed through the wall! At the age of 50, Röntgen had discovered a new ray. He called it "X ray," meaning "unknown ray"—a name that has stuck, even though his ray is no longer so mysterious.

THE BEGINNING OF MODERN PHYSICS

In the 1960s, Richard Feynman, an American theoretical physicist, used to ask his students, "What do we mean by 'understanding the world?'" Then he'd tell his packed lecture halls at the California Institute of Technology that trying to understand the world is a lot like watching a giant chess game for which you do not know the rules—and trying to figure out the rules from what you see. Eventually you may catch on to a few of the rules, and *"the rules of the game,"* he said, "are what we mean by *fundamental physics."* But most situations are so complex that, even if we knew every rule, we "cannot follow the plays of the game using the rules, much less tell what is going to happen next."

As discoveries began to pile up in the late 1890s and the early years of the 20th century, physicists began to get their first real taste of the complexities that Feynman would present to his students more than a half-century later.

Wilhelm Röntgen probably had no idea that his discovery would touch off a revolution in physics, providing an event that scientists would point to as a watershed from the old, "classical" physics to the modern age. From 1895 on, physics would never be the same again.

These were exciting years for physicists—from 1895 to 1945—perhaps both the most exhilarating and confusing 50-year span in the history of science. The 2,300-year-old concept of the atom would undergo dramatic metamorphoses during these years. What once had been thought of as the ultimate tiny, unsplittable particle would slowly become transformed in people's minds, as physicists discovered its even tinier parts: negatively charged electrons spinning like submicroscopic planets around a compact nucleus. The nucleus, too, held surprises: positively charged proton particles and neutron particles that had mass, but no charge. The discovery of other, even tinier, particles would follow. What's more, by 1945, it was clear not only that the atom was composed of even tinier particles, but that atoms, unquestionably, could be split.

New concepts about space and time would shake the very principles of physics, which had been established so firmly by Isaac Newton in the 17th century. And a new, totally baffling idea called quantum theory would subvert logic while laying the groundwork for nearly every major techno-logical advance surrounding us in our world today—from television sets to digital watches.

From Röntgen's X rays, two big boulders started rolling. One would begin an avalanche of revolutionary new ideas about the atom; the other would lead to the discovery of a strange instability in certain elements, a characteristic that would enable us ultimately to tap into nuclear power. But we're getting ahead of ourselves. At the time Röntgen discovered X rays, the idea of an atomic "nucleus" did not even exist.

The idea of the atom goes back to the 5th century B.C., when a Greek philosopher named Leucippus and his student Democritus said that all matter was made up of tiny, indivisible particles. (The word *atom* comes from the Greek word *atomos*, which means "uncuttable" or "indivisible.") An atom, by definition, was the smallest possible particle of matter. (A diagram of Leucippus's atom would look like an extremely small billiard ball.) Not everyone liked the idea; in fact, Plato and Aristotle, two enormously influential Greek thinkers, didn't take to it at all. As a consequence, with the exception of a few renegade atomists (including Newton), the idea lay more or less dormant for about 2,200 years, until John Dalton resurrected it in the early 19th century with his "atomic theory." Dalton thought that all matter consisted of atoms—and by this time other scientists began to become interested in the idea because Dalton was able to show quantitative scientific evidence for atoms. He also showed how this theory fit with two ideas that finally had begun to catch hold: the law of the conservation of matter (set forth by Galileo) and the law of constant composition, or law of definite proportions, formulated by Dalton. By Röntgen's time, the atom had become well accepted as the ultimate indivisible particle of nature—but everyone still thought of it as a sort of ultimately tiny billiard ball.

THE NEW RAY

Röntgen delayed seven weeks before announcing his exciting discovery. He wanted to be sure. (Years later, one inquirer asked him what he had thought when he made his discovery, to which he replied pithily, "I didn't think; I experimented.") When he did announce the new rays on December 28, 1895, he had all the pertinent details in hand, including the fact that X rays not only penetrate opaque matter but they can also impart an electrical charge to a gas and are not influenced by magnetic or electric fields. The world was stunned by his discovery, and physicists were amazed.

People immediately saw the potential for the use of X rays for medical diagnosis (although, unfortunately, it was not until many years later that they discovered that X rays could also be dangerous). X rays could pass easily through soft-body tissue, while being largely blocked by bone structures or

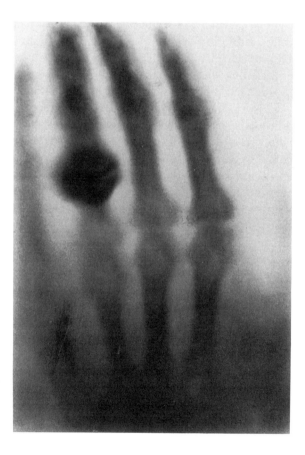

An X-ray photograph of the hand of Röntgen's wife
(National Library of Medicine)

Henri Becquerel discovered radioactivity in 1896 (Burndy Library, Courtesy AIP Emilio Segrè Visual Archives)

other more solid materials. So if a photographic plate is placed behind a patient, a photo can be taken showing bones as a white shadow on black. Tooth decay looks gray against the white of the teeth. Metal objects also show up clearly, and within four days after Röntgen's news arrived in America, X rays were used to locate a bullet lodged in a patient's leg. Just three months after Röntgen's announcement, a boy named Eddie McCarthy in Dartmouth, Maine became the first person to have a broken bone set using the new way to view bones.

Röntgen had caused a great furor, not entirely positive. In the state of New Jersey, legislators worried that X rays meant the end of personal privacy (they were particularly concerned about the modesty of young women) and proposed legislation to prevent the use of X rays in opera glasses—an unnecessary worry, of course.

But for scientists Röntgen's X rays (initially known as Röntgen Rays) would become one of the greatest tools in biological research, and their discovery marked the beginning of a second scientific revolution in physics. For his discovery, in 1901 Wilhelm Röntgen became the first person ever awarded the Nobel prize in physics.

Meanwhile, in the spring of 1896, in Paris, a physicist named Henri Becquerel became fascinated with the new rays.

URANIUM'S STRANGE GIFT

Henri Becquerel (1852–1908) [beh KREL] heard about Röntgen's X rays on January 20, 1896. By February he had begun to experiment. Becquerel's family had been physicists for several generations, and both he and his father before him had done work on fluorescence, the tendency of some materials to absorb radiation and then glow. Since Röntgen had discovered X rays from the fluorescence they caused, Becquerel wondered if the reverse might also be true. Could fluorescing bodies, he wondered, produce X rays or cathode rays? (Cathode rays are streams of electrons projected from the surface of a cathode, or the negative electrode of an electrolytic cell, or battery.)

So he devised a test on certain salts of uranium that he knew gave off a fluorescence when exposed to sunlight. In February 1896 Becquerel took a crystal of fluorescent uranium salts and placed it in the sunlight on top of some photographic paper wrapped in black paper. The idea was that if the fluorescent salt emitted X rays, those rays would penetrate the black paper and expose the film. Otherwise, no light he knew of, including ultraviolet light, could reach the film through the black paper.

When Becquerel developed the photographic film, it was indeed fogged, as if exposed to a glow of light. Aha, thought Becquerel: Fluorescence produces X rays.

Of course, one trial doesn't prove a point in science, and Becquerel planned several other experiments to test his hypothesis. But a series of cloudy winter days in Paris hindered his plans. So he set the salts aside, neatly stacked on top of an unexposed photographic plate and shut away in a dark laboratory drawer, waiting for a sunny day. Finally, impatient with the weather, on March 1 he took out the plate to develop it anyway. He thought some fluorescence from the original exposure might have remained in the crystals. Maybe the plate would show at least a little fogging as a result. And this is where the big surprise came in. The plate showed not the slight, dim fogging he thought might be there, but a strongly visible clouding. No sunlight could have penetrated the darkened drawer and the careful wrapping, and no leftover fluorescence from a previous exposure could have caused this effect. Becquerel had discovered a new kind of radiation that no one had ever known about before!

Excited, he began studying this new radiation and discovered that it had several characteristics in common with X rays. It could penetrate opaque matter. It ionized air (gave it an electrical charge). And it was emitted by an object in all directions, in a constant stream. Stimulated by Röntgen's discoveries, Becquerel had made—purely by accident, as often happens in science—another, equally exciting discovery. One, as it happened, that would produce a long line of fruitful investigations for many years to come.

*Marie Curie,
discoverer of polonium
and radium* (AIP
Emilio Segrè Visual
Archives)

THE CURIES' QUEST

Several of those investigations would be conducted by an intense young woman who had arrived in town only four years before. When 25-year-old Marie Sklodowska reached Paris from her native Warsaw in 1892, she could not have dreamed to what pinnacles her intelligence and unparalleled determination would take her—that she would not only become the first woman to win a Nobel prize, but that she would earn the rare distinction of winning two of them, one in physics and one in chemistry. She couldn't have known, either, that she would find a husband, Pierre Curie, who was a perfect match for her in every way: a distinguished fellow scientist who shared her passion for knowledge and achievement, who loved her and whom she loved, and who would labor at her side with equal dedication, often as her assistant. Nor could she have envisioned—and in fact she never knew—that she would begin a two-generation dynasty of Nobel laureates. (Her daughter, Irène Joliot-Curie, would also win a Nobel prize, with her husband Frédéric Joliot-Curie, in 1935, the year after Marie's death.) But she knew that she finally had obtained the opportunity she had been working toward for years: entrance to the Sorbonne.

Life in Poland in the 1890s was repressed, under the domination of Russia, and Marie had already helped send a brother and a sister away for a better education at the great Parisian university. Now it was her turn. Even so, she was living on meager rations—in fact, she fainted once in class, from hunger. But Marie Sklodowska graduated at the top of her class.

In 1894 she met Pierre Curie, a well-established French chemist already known for his work in the field of piezoelectricity. The two were married on July 25, 1895 in a civil ceremony. Always frugal, both then and later in life, they set off on their honeymoon in southern France on bicycles.

By the end of that same year, Röntgen announced his discovery of X rays, and only a few months later the news of Becquerel's discovery swept through Paris and the worldwide physics community. In England, J. J. Thomson at the Cavendish Laboratory jumped at the news and easily persuaded one of his young students, 24-year-old Ernest Rutherford from New Zealand, to turn his attention to X rays.

"The great object is to find the theory of the matter before anyone else," Rutherford wrote home to his family, "for nearly every professor in Europe is now on the warpath." The same was shortly also true of the radiation Becquerel had discovered. The question was: What could be causing these unexplained radiations and of what were they composed?

Marie Curie leaped into this exciting new field. She soon discovered—at roughly the same time that Becquerel and Rutherford did—that the radiations given off by uranium were composed of more than one type. Some rays were bent one way by a magnetic field; others were bent another way. Rutherford named the positively charged rays alpha rays and the negatively charged ones beta rays (also known as alpha particles and beta particles). Exactly what these rays or particles were composed of, no one knew, but by 1898 Marie Curie suggested a name for these radiations—*radioactivity*—and that is the name that stuck. And in 1900, Paul Ulrich Villard discovered a third, unusually penetrating type of ray in radioactive radiation, one that did not bend at all in a magnetic field, which he named the gamma ray. (The use of Greek letters to name these rays simply meant that their identity was unknown, as with the X in X ray.)

Marie Curie, meanwhile, used a discovery of her husband Pierre's to measure radioactivity. Radioactive rays, like X rays, ionized any gas they passed through (including air) making it capable of conducting electricity. She found that she could measure the current so conducted with a galvanometer and offset it with the potential of a crystal under pressure. By measuring the amount of pressure it took to balance the current, she could obtain a reading of the intensity of the radioactivity. She systematically tested radioactive salts and succeeded in showing that the degree of radioactivity was in proportion to the amount of uranium in the radioactive material—thereby narrowing the source of the radioactivity in her samples down to uranium. Then in 1898 she made yet another find: the heavy element thorium was also radioactive.

Even more interesting is the fact that as Marie was working to separate uranium out of pitchblende, she found that the residues she produced had a much higher measurement of radioactivity than the uranium content alone could account for. Since the other minerals present in the ore were not radioactive, that could mean only one thing: Some other radioactive element, in amounts too small to detect, must also be present!

By this time, Marie's work had developed so much potential that her husband Pierre joined her to help with the backbreaking, tedious work of crystallizing the elements from the ores. Though himself a fine scientist with a successful career, he set his own work aside and spent the remaining seven years of his life assisting her, recognizing both her extraordinary gifts as a scientist and the importance of the path she was following. (In 1906, two years after receiving an appointment as professor of physics at the Sorbonne, he was run over by a horse-drawn dray at the age of 47. Marie, appointed in his place, would become the first woman to teach physics at the Sorbonne.)

By July 1898 the two had succeeded. Working together, they had isolated a tiny amount of powder from the uranium ore. It was a new element, never before detected, with a level of radioactivity hundreds of times higher than uranium. They named the new element polonium, after Marie's home country.

But something still seemed strange. The ore still gave off more radioactivity even than the uranium and polonium combined could account for. There must still be something else. In December 1898 they found the answer: another, even more radioactive element. This one they named radium.

But radium was practically a ghost. They could not really offer a good description of this new element because the amount they were able to derive from the ore they had was so minuscule. They could measure its radiations, and Eugène Demarçay, a specialist in elemental line spectra, was able to provide the spectral characteristics.* But that was all.

So the next project was to produce a large enough quantity of radium that they could weigh it and measure it and see it. They spent their life savings to obtain large masses of waste ore from a nearby mine, and they began the monumental task. They spent four years, during which Marie lost 15 pounds, purifying and repurifying the ore into small amounts of radium. It took eight tons of pitchblende to produce one gram of radium salt.

Marie Curie wrote her doctoral dissertation on the subject in 1903, for which she, Pierre and Henri Becquerel shared the Nobel prize in physics

---

* Different elements give off different wavelengths of electromagnetic radiation or light, and these can be observed as discrete lines. (See figure on page 12.)

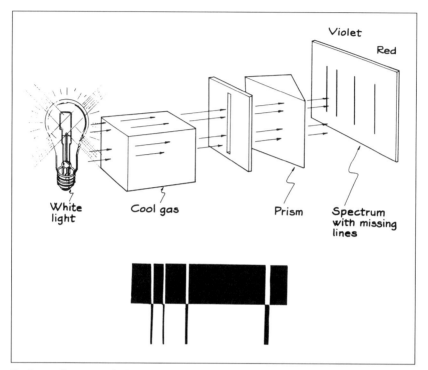

In the top illustration, light emitted by an incandescent bulb produces an absorption spectrum when it passes through a sample of cool gas before striking a screen. The gas has absorbed energy from the white light of the bulb, as indicated by the dark lines on the screen. From the pattern of dark lines it is possible to tell which frequencies the gas has absorbed and, because every element has a unique spectrum, the gas can be identified from the pattern. When an element is heated, it gives off (emits) energy that can be displayed in the same way, producing a similar emission spectrum. The bottom illustration shows the emission and absorption spectra of hydrogen. (Jacqueline D. Spears and Dean Zollman, *The Fascination of Physics*. Benjamin/Cummings, 1985.)

that year. Eight years later, after Pierre's death, she received another for her discovery of two new elements, this time in chemistry and this time alone, her partner no longer at her side.

AMAZING ELECTRONS

Meanwhile, as the Curies labored to crystallize tiny amounts of rare, previously unknown elements, J. J. Thomson, working at Cambridge in England, also became immediately fascinated with Röntgen's X rays.

Joseph John Thomson (1856–1940), a son of a bookseller, was so brilliant that he entered the University of Manchester at the age of 14. There he planned to study engineering, but he dropped that career plan when his family didn't have the money to finance further studies. So he became a physicist instead—luckily for the development of modern physics. In 1876 he earned a scholarship to Cambridge University, which became his home for the rest of his life. Seven years later he became professor of physics there, and by 1884 he was director of the Cavendish Laboratory, where he stimulated the minds of an entire generation of young scientists until his retirement in 1919. Thanks to his influence the Cavendish became the premier hotbed for atomic studies for the next three decades.

Thomson initially became interested in the theories of James Clerk [CLARK] Maxwell (1831–79) on electromagnetic fields, and he became intrigued, as Röntgen initially had been, by the cathode ray, since it appeared not to be electromagnetic in nature. Other people had shown that a cathode ray could be deflected, or turned aside, by a magnetic field. This, they said, showed that cathode rays were composed of negatively charged particles. No one had ever succeeded in showing that a cathode ray could be deflected by an electrical field; yet if cathode rays were charged particles, an electrical field should deflect them.

Thomson took up the challenge in 1896. Working with a cathode ray tube, he introduced an electrical field and succeeded in deflecting the rays. He measured the deflection and the velocity. He tried different cathode materials: aluminum, copper, tin, platinum. He tried different gases in the tube: air, hydrogen, carbon dioxide. All the figures remained the same. He realized that if cathode rays were simply charged atoms, as some other researchers had thought, then the numbers would be different, reflecting the different masses of the atoms.

By 1897 he was satisfied that he had isolated what he called a "negative corpuscle," which he believed must be a fundamental particle—smaller than the atom—that was a part of all matter. Thomson had discovered that the atom had parts! "We have in the cathode rays matter in a new state," he announced, "a state in which the subdivision of matter is carried very much further than in the ordinary gaseous state." He called it a "subdivision of matter" that is part of the "substance from which the chemical elements are built up." He borrowed the term *electron* from the term coined by Irish physicist George Johnstone Stoney in 1891 for the unit of electricity that is lost when an atom becomes an ion (a charged particle).

This was a stunning piece of news in the world of physics and chemistry. For the first time, scientists began to suspect that the atom had an internal structure. The very identity of the atom—that basic constituent of nature— was threatened. Hydrogen had long been clearly established as the lightest of all atoms, and yet here was a particle that was 2,000 times lighter than hydrogen. The only possible explanation was that this new negative corpuscle was a subatomic particle—a basic building block even more fundamental than the most fundamental. It was the only possible explanation, and yet it was impossible in terms of every tenet of the old atomic theory. As one science historian put it, "The deconstruction of the atom into subatomic particles had begun." It would prove to be a long road, and even today we may not have reached the end.

Meanwhile, in France, in 1899 Henri Becquerel noticed that the radioactive radiation he had discovered could also be deflected by a magnetic field. From this he deduced that at least part of the rays that he had been studying were also tiny, electrically charged particles. And by 1900 he had come to the conclusion that the negatively charged particles in radioactive radiation were identical to the electrons that Thomson had discovered in cathode rays.

By 1901 Becquerel had also recognized that the uranium portion of the salts he'd been working with was the source of the radiation Marie Curie had called radioactivity, and he concluded that the only place the electrons could be coming from was the uranium atom itself.

What effect did all this have, then, on the definition of atoms? Obviously, they weren't smooth, featureless miniature billiard balls, as Dalton had imagined. And they clearly weren't indivisible.

PLUM PUDDING

J. J. Thomson immediately saw that some new model of the atom was in order. Electrons had a negative charge; yet matter had no charge. So the atom must have some internal structure that offset the negative charge of electrons with a positive charge. In 1898 Thomson proposed what has come to be called the "plum pudding" or "raisins-in-poundcake" model of the atom: negatively charged electrons embedded in a uniform, positively charged sphere of matter.

Thomson, who was notoriously clumsy in the lab, is sometimes overshadowed by others who followed. (His own son, George, said, "though he could diagnose the faults of an apparatus with uncanny accuracy it was just as well not to let him handle it.") But he left a great legacy. He blazed the trail to atomic particle physics, and he won the Nobel prize in physics in 1906 for his work on the electron. Seven of Thomson's research students later went on to win the Nobel Prize, including his son George, who would later demonstrate the wave nature of the electron. By the time Thomson reached his fifties and sixties, from about 1906 to 1919, he had developed a reputation as the "great man" of Cambridge. As one student later described him, however, "he was not fossilized at all," but was friendly, "quite youthful

Ernest Rutherford, about 1906 (AIP Emilio Segrè Visual Archives)

still"—and shaved rather badly. That student was Thomson's first, and perhaps most famous, research student, an exuberant young man with big hands and a big walrus mustache: Ernest Rutherford.

THE CROCODILE FROM NEW ZEALAND

When Ernest Rutherford (1871–1937) arrived on scholarship at the Cavendish Laboratory from his native New Zealand, he was 24 years old, a large, dark-haired man with strong opinions, plenty of ambition and no money. The competition was keen, he wrote home to his mother. "Among so many scientific bugs knocking about, one has a little difficulty in rising to the front." But Rutherford was never one to let anything, much less a little competition, stop him. In later years, his students nicknamed him "the crocodile" because, as one explained, "the crocodile cannot turn its head . . . it must always go forward with all devouring jaws." He was tireless in every quest and loved his role as one who put endless questions to nature. His great success, as his colleague Niels Bohr once said, came from "his intuition in shaping such questions so as to permit the most useful answers."

He was a man who "formulated hypothesis after hypothesis," as Italian-American physicist Emilio Segrè [SAY gray] once described him, "rejecting them or modifying them according to need, doing everything with inexhaustible energy. He worked all the time, and even his friends and colleagues barely knew a small fraction of his scientific thoughts."

A consummate experimentalist, Rutherford generally had little use for theorists. "They play games with their symbols," he said, "but we [experimentalists] must turn out the real facts of Nature." He had a particular talent for designing experiments and an uncanny ability to pick out one significant fact from a mass of confusing detail. As one colleague remarked, "With one movement from afar, Rutherford so to speak threaded the needle the first time."

In 1898, Rutherford accepted an appointment as professor of physics at McGill University in Canada, and while there, from about 1902 on, he and his assistants began experimenting with alpha particles to find out more about them. By 1908–09, Rutherford had returned to England, to the University of Manchester, where a young German physicist named Hans Geiger teamed up with him. Together, they bombarded thin pieces of gold foil with alpha particles. Most of the alpha bombarders passed right through the foil, which was exactly what the experimenters expected, based on Thomson's model of the atom. But some of the alpha particles struck the

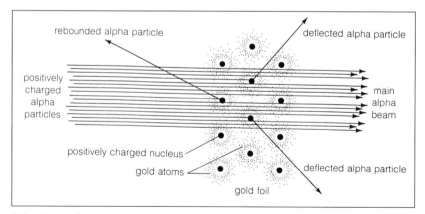

When Rutherford shot positively charged alpha particles at a piece of gold foil, most of the alpha particles passed straight through. But, to Rutherford's surprise, a few were deflected sharply, revealing the presence of the gold atoms' small, positively charged nuclei, whose existence no one had suspected. (Jacqueline D. Spears and Dean Zollman, *The Fascination of Physics.* Benjamin/Cummings, 1985.)

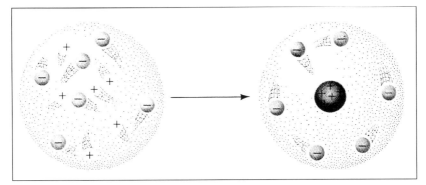

Models of the atom: Based on his discovery of the electron, J. J. Thomson suggested in 1898 (left) that atoms were spheres of positively charged matter with negatively charged electrons embedded in them—something like raisins in poundcake. By 1911, his student Ernest Rutherford had come up with the idea that atoms consisted of a tiny positive nucleus with electrons circling somewhere outside it. (G. Tyler Miller, Jr., Chemistry: A Basic Introduction, 2nd Edition. Wadsworth Publishing Company, 1981.)

gold foil and were deflected at a sharp angle—often 90° or more. This amazed Rutherford, who remarked, "It was as though you fired a 15-inch shell at a piece of tissue paper and it came back and hit you."

Early in 1911 Rutherford came buoyantly to Geiger. "I know what the atom looks like!" he exclaimed. Based on his results, Rutherford had put together a new idea of the atom: What if all the positively charged particles in the atom were not spread like a fluid throughout the atom as Thomson had thought, but were lumped together in the center in one tiny area, or "nucleus"? Most of the atom's mass would be contained in the nucleus, and an equal number of negatively charged electrons would be found in motion somewhere outside the nucleus. It was a compelling idea—a sort of tiny planetary system that echoed the larger Solar System we live in.

Rutherford's idea of an atomic nucleus was a zinger, one for which he has earned the title "the Newton of atomic physics." It seemed to solve all the problems with the raisins-in-poundcake model for atoms. Yet even this model had a few problems. To build a more accurate vision of the nature of the atom would require the application of an amazing concept called "the quantum" set forth by a somewhat dour German scientist named Max Planck. Like Röntgen's X rays, this idea would virtually turn physics upside-down, with implications not just for the concept of atoms, but for virtually everything about our understanding of how the world works.

THE NEW UNIVERSE,
PART ONE:
EINSTEIN AND RELATIVITY

*E*xciting as the new atomic physics was, the world of physics was changing in broader, even more far-reaching ways, with ramifications for our understanding of the very shape of time, space and the universe. This part of the second scientific revolution was the work of Albert Einstein, a brilliant and creative theorist and the only thinker ever to be ranked in the same class as Newton. But for this part of the story, we need to go back to James Clerk Maxwell and his ideas about light.

Maxwell had introduced a revolutionary set of equations that verified the existence of electromagnetic fields and established that magnetism, electricity and light were a part of the same spectrum: the electromagnetic spectrum. Light, he maintained, was a wave, not a particle, and he thought that it traveled through an invisible medium he called "the ether," which filled all space. But physicists began to see a problem, not with Maxwell's electromagnetic field equations, but with his ideas about the ether.

THE ETHER PROBLEM

Maxwell wasn't the first to come up with this idea that some invisible medium called the ether must fill the vastness of space, extending "unbroken from star to star." It dated back to the time of the ancient Greeks. "There can be no doubt," Maxwell said in a lecture in 1873, "that the interplanetary and interstellar spaces are not empty but are occupied by a material substance or body, which is certainly the largest, and probably the most uniform, body of which we have any knowledge."

Teamed up with Edward Morley, Albert Michelson showed that the "ether" that James Clerk Maxwell thought light traveled through does not exist. (Astronautical Society of the Pacific)

The idea of the ether seemed necessary because, if light was a wave, it seemed obvious that it had to be a wave traveling in some medium. But accepting what "seems obvious" is not the way to do good science; if the ether existed, it should be possible to find some proof of its existence.

Albert Michelson (1852–1931), an American physicist, had an idea. If the ether that filled the universe were stationary, then the planet Earth would meet resistance as it moved through the ether, creating a current, a sort of "wind," in the ether. So it followed that a light beam moving with the current ought to be carried along by it, whereas a light beam traveling against the current should be slowed. While studying with Hermann von Helmholtz in Germany, in 1881 Michelson [MY kul sun] built an instrument called an interferometer, which could split a beam of light, running the two halves perpendicular to each other, and then rejoin the split beam in a way that made it possible to measure differences in the speeds with great precision.

Michelson ran his experiment, but he was puzzled by his results. They showed no differences in light velocity for the two halves of the light beam. He concluded, "The result of the hypothesis of a stationary aether is . . . shown to be incorrect, and the necessary conclusion follows that the hypothesis is erroneous."

But maybe his results were wrong. He tried his experiment again and again, each time trying to correct for any possible error. Finally, in 1887, joined by Edward Morley, Michelson tried a test in Cleveland, Ohio. Using improved equipment, and taking every imaginable precaution against inaccuracy, this time surely they would succeed in detecting the ether. But the experiment failed again.

As a result, the Michelson-Morley experiment has become the most famous failed experiment in the history of science. They had started out to study the ether, only to conclude that the ether did not exist. But if this were true, how could light move in "waves" without a medium to carry it? What's more, the experiment indicated that the velocity of light is always constant.

It was a completely unexpected conclusion. But the experiment was meticulous and the results irrefutable. Lord Kelvin, one of the great gurus of physics of the time, said in a lecture in 1900 at the Royal Institution that Michelson and Morley's experiment had been "carried out with most searching care to secure a trustworthy result," casting "a nineteenth-century cloud over the dynamic theory of light."

The conclusion troubled physicists everywhere, though. Apparently, they were wrong about the existence of the ether—and if they were wrong, then light was a wave that somehow could travel without a medium to travel through.

What's more, the Michelson-Morley results seemed to call into question the kind of Newtonian relativity that had been around for a couple of centuries and by this time was well tested: the idea that the speed of an object can differ, depending upon the reference frame of the observer. Suppose two cars are traveling along on the freeway. (There weren't many cars or freeways in 1887, but you get the idea.) One car is going 55 miles per hour, the other 54 miles per hour. To the driver of the slower car, the faster car would be gaining ground at a rate of 1 mile per hour. Why would light be any different?

But that's just what the Michelson and Morley experiment had shown: Light does behave differently. The velocity of light is always constant—no matter what. Astronauts traveling in their spaceship at a speed of 185,000 miles per second alongside a beam of light (which travels at 186,000 miles per second) would not perceive the light gaining on them by 1,000 miles per second. They would see light traveling at a constant 186,000 miles per second. The speed of light is a universal absolute.

They had uncovered what one science writer calls a "deep enigma of nature." Less than five years later, though, one of the greatest scientists of all time would grab hold of the idea of the constant speed of light and begin to make sense of the chaos with his special theory of relativity.

But first, yet another revolution was brewing.

THE QUANTUM MYSTERY

Max Planck wasn't the type to start a revolution. Tall and spare, quiet and dignified—some said "stuffy and pedantic"—he was completely devoted to tradition and authority, both in his physics and in his life. Born in Kiel, Germany in 1858, Planck was not a remarkable child. His father was a professor of civil law and the family moved to Munich, where young Max began his education when he was nine years old. His childhood was uneventful and so were his college years. Studying physics in Berlin, he took classes from such respected German scientists as Hermann von Helmholtz, Rudolf Clausius and Gustav Kirchhoff, and his work was competent but

Max Planck, about 1930 (Max Planck Gesellschaft, Courtesy AIP Emilio Segrè Visual Archives)

promised nothing special. For his doctoral work he chose thermodynamics because he admired Clausius's work in the same field, but there was little in his dissertation to excite anyone's attention; to all outward indications he was headed straight for a largely uninspired and mediocre career. Even after securing a professorial appointment at the University of Berlin in 1889, Planck appeared to be traveling pretty much on the beaten path. Certainly his choice of thermodynamics as his specialty didn't promise much in the way of major achievements. In fact, when he first started his scientific work in college, one of his professors warned him against pursuing physics because the field was heading for a dead end—since all the great work had already been done and all that was left for new physicists to accomplish was the tiresome task of clearing up a few minor details.

It so happened that one of the "minor details" left to tidy up in the field of thermodynamics was known as the *ultraviolet catastrophe*, a name dramatic enough to attract anyone's attention. It was a result of problems involved in trying to understand what had come to be referred to as blackbody radiation.

A blackbody, in physics, is one that absorbs all frequencies of light and reflects not even the slightest little shimmer. A piece of coal comes close. So does a toaster wire. Logically, since a blackbody absorbs all frequencies, it should, when heated, radiate all frequencies as well. But here's the rub. Physicists expected the number of frequencies in the high range to be far greater than the number of frequencies in the low range—since high frequencies have shorter wave lengths and more could be packed into the black body. So the problem with blackbody radiation is that, if a body radiates all frequencies equally, the number of radiations in the high-frequency range should vastly outnumber those in the low-frequency range. In fact, almost all the radiation should be high-frequency—that is, at the ultraviolet end of the spectrum.

But that doesn't happen. And no one could explain why in terms of physical theory in the 1890s, although several people certainly had tried.

As the physicist Emilio Segrè wrote in 1980 in his study of modern physics, *From X-Rays to Quarks*, "Planck's love of fundamental and general problems drove him to take up the blackbody problem, which was independent of atomic models or other particular hypotheses. He loved the absolute, and such was the blackbody."

He might have been conservative, and he might have been stuffy, but Max Planck also wanted to be the best at anything he did, even if he never tried to do any really big things. He gave up studying piano as a child because he thought he could never be a *great* pianist, only a good one. The blackbody problem appealed to him because it was containable, it didn't stretch all over the place—it was like a complicated jigsaw puzzle. The pieces were all there, he was sure, even if they were all in a jumble, and all he had to do was dump them on a table, sort them out and put them together in the right way. No one else had done it. But he was certain he could. And then a small piece of immortality would be his.

He spent more than six years of his life trying to find a solution. And after he found it, physics would never be the same again. For, in quietly solving the puzzle of blackbody radiation, Max Planck would discover a key principle that, followed up by others, would forever change our understanding of the world.

As one science historian put it, "Planck was like a man who, before the discovery of fire, wanted to find the best ways to bore holes, who for months, years, even decades, bored holes in every material he could find in every conceivable fashion and in doing so chanced to discover fire."

Based on a hunch, in 1900 Planck worked out a simple equation describing the distribution of radiation accurately over the entire stretch of frequency. His basic premise went like this: What if energy is not infinitely indivisible? What if, like matter, energy exists in particles or packets, or *quanta*, as he called them—from the Latin interrogative adjective meaning "How much?" (Quanta is the plural; the singular form is quantum.)

Planck also found that the size of these quanta was in direct proportion to the frequency. So radiation at low frequencies is easy—it requires only small packets or quanta of energy. But for a frequency twice as high, radiation would require twice the amount of energy.

Following Planck's idea that energy can only be emitted in whole quanta, it becomes very easy for a body to radiate at low frequencies: not that much energy has to be pulled together to make up a quantum of energy. But at high frequencies, pulling a quantum's-worth of energy together is not so easy. The quantum-energy requirements to radiate at the high-frequency end of the spectrum are so great that it's very unlikely to happen. So blackbodies do not radiate all frequencies equally—and that's the key to the so-called ultraviolet catastrophe.

As temperature goes higher, it becomes easier for the larger quanta of energy to form that are required for radiation at higher frequencies—and so radiation at those frequencies becomes more likely. That's why a minimally heated body (for example, the human body) radiates only in the infrared spectrum. A bar of iron heated to relatively low temperature glows red, but as it's heated to a higher temperature, its color changes, first to orange, then to yellow, then to blue. If it were heated high enough, it would glow white, reflecting all frequencies at once.

The ratio of the frequency of the radiation and the size of the quantum is a constant, h, known as Planck's constant and now recognized as one of the fundamental constants of the universe.

Planck had solved the blackbody puzzle, but once it was laid out before him with all of its full implications he wasn't happy with the final picture he saw. He didn't want to see classical physics destroyed, and the quantum theory would do that. Still, he knew that he had started something that could not be stopped. The power of the theory was too great, even if its implications were frightening to him. "We have to live with quantum theory," he concluded, "And believe me, it will expand. . . . It will go in all fields."

Max Planck, though, was not the one who would push it into those fields.

For the rest of his life he would be famous for discovering the quantum, but he would spend his working days trying somehow to reconcile his disturbing discovery with his beloved classical physics. It was an attempt bound to fail. "My vain attempts to somehow reconcile the elementary quantum with classical theory continued for many years, and cost me great effort," he wrote near the end of his life.

Many of my colleagues saw almost a tragedy in this, but I saw it differently because the profound clarification of my thoughts I derived from this work had great value to me. Now I know for certain that the quantum of action has a much more fundamental significance than I originally suspected.

Among the first to recognize the fundamental significance of the quantum was a fellow German who would use Planck's theory to solve another disturbing mystery of physics, who would then proceed to shake up science with some independent and revolutionary ideas of his own, and would finally return to the mysteries of the quantum and, like Planck, spend the remainder of his life locked in battle with what the quantum had wrought.

EINSTEIN AND THE PHOTOELECTRIC EFFECT

If Max Planck was noted for his middle-class conservatism, Albert Einstein (1879–1955) was the epitome of the complete rebel: a man who preferred to work alone, a wanderer in the highest realms of thought, uncomfortable with and often disdainful of the everyday preoccupations of the average man or woman. As he explained once, he had sought out science to get away from the "I" and the "We." He preferred to think instead about the "It." As a wise and "grandfatherly" elder in his later years, he was the "saintly," eccentric and lovable genius personified. In his youth, though, and through the period of his greatest achievements, he could be curt, impatient, egotistical and selfish. Supremely confident of his own genius, like his greatest predecessor, Isaac Newton, Einstein walked his own way, set his own challenges, and worried little about what he called "the chains of the merely personal . . . dominated by wishes, hopes, and primitive feelings."

Albert Einstein was born in Ulm, Germany in 1879, the same year that the greatest theoretical physicist of the 19th century, James Clerk Maxwell,

Albert Einstein   (Mount Wilson Observatory)

died. Einstein was the "odd man out" even in his childhood and early school years. A precocious child, he was also lonely and sometimes bitter. Preferring to teach himself, he hated the regimentation and pedantic methods of the German school system. "When I was in the seventh grade at the Luitpold Gymnasium (and thus about fifteen)," he later wrote in a letter, "I was summoned by my home-room teacher who expressed the wish that I leave the school. To my remark that I had done nothing amiss he replied only 'your mere presence spoils the respect of the class for me.'

"I myself, to be sure, wanted to leave school . . . the main reason was the dull, mechanized method of teaching. Because of my poor memory for words, this presented me with great difficulties that it seemed senseless for me to overcome. I preferred, therefore, to endure all sorts of punishments rather than learn to gabble by rote."

In 1894, Albert Einstein dropped out of school, after talking his family doctor into giving him a certificate stating that he needed rest and recuperation for health reasons. He was 15 years old and he had a plan. His intention was to skip the regimentation of the gymnasium, spend a year in wandering and self-study, and then take the entrance examination at the famous Federal Institute of Technology in Zurich, Switzerland. As he often remembered it later, the following year was one of the happiest times of his life. He spent it tramping around the mountains of Germany and Italy, studying his physics books and visiting art galleries and museums in Genoa. His plan failed, however. At age 16 he took the entrance exams at the "Poly," as the institute was affectionately known, and he failed them.

The situation might have been a disaster, and young Einstein might have fallen into painful obscurity like many other school dropouts, but he had done well enough in math and physics to capture the attention of one of the professors, who encouraged him to sit in on his physics lectures, anyway. He was also urged not to give up hope, but to apply for admission at the progressive Swiss Cantonal school.

Blossoming in the freedom and lack of regimentation of the Swiss school, he quickly obtained his diploma and once more applied for admission to the "Poly." This time, with the age requirements waived for him, he was accepted in the fall of 1896.

Four years later, in 1900, he obtained his degree from the institute. The years, though, had not been happy ones. He loved Switzerland, so much so that he obtained a Swiss citizenship in 1901, but he was just not a happy student. Even the relative freedom of the Zurich Institute was too confining for his temperament. As he later wrote about his years at the institute in his *Autobiographical Notes*, "One had to cram all this stuff into one's mind for the examinations whether one liked it or not. This coercion had such a deterring effect on me that, after I passed the final examinations, I found the consideration of any scientific problems distasteful for me for an entire year."

Things didn't get any better after graduation. He hadn't made many friends or connections at the institute and had in fact alienated some of his professors (who might have helped him obtain a university position) with his disdainful attitude. His financial situation was precarious and no positions appeared on the horizon. Living hand-to-mouth over the next year, he pieced together enough temporary positions, tutoring and acting as occasional substitute teacher, to keep going. With his love for science slowly returning after his disheartening years at the institute, he picked up his studies again and began to research for his doctorate.

Finally, he had a bit of luck. And like most things about Albert Einstein, it was an unconventional and unexpected bit of luck. One of the few professors he had not managed to alienate at the institute knew that Einstein needed a job and spoke to his father about him. The father then recommended Einstein to his friend Friedrich Haller, the director of the Swiss Patent Office in Bern. There was a position open and Einstein was interviewed. For someone like Einstein, the job was easy. It consisted of looking over the new patent applications as they came in and deciding if they were scientifically or technically feasible before passing them on to a higher official. Einstein's job interview went well enough, but the job was a civil service position, and according to law the position had to be advertised. Einstein was told that he would be notified after other applicants had been screened. The final decision would take months. It was a tough wait, but Einstein spent some of the time preparing an article on thermodynamics, which he submitted to the "Poly" as a thesis and a partial requirement toward his doctorate. The article was eventually rejected as a thesis (although he later would find a publisher for it), and things seemed to be continuing downhill for him, when he was finally notified that the position at the patent office was his.

In June 1902, he went to work at the Swiss Patent Office. It was a far cry from the academic world, and Einstein thought the job was perfect. The rebel had found a temporary home. He was free from the pedantic thinking and the rigid academic regimentation that he hated so much and he was pretty much left on his own. The work was interesting, easy and often amusing, especially when the more bizarre plans for crackpot scientific inventions passed across his desk. He had a good scientific eye for what was right and what was wrong with a proposed invention, and he was perfectly happy to send the more intelligent and original concepts on to his superiors. More important, though, he usually could finish up his entire day's work in a few hours in the morning, leaving the rest of his office day free to think and work on his own scientific ideas, for which he luckily required no laboratory—just a sharpened pencil, a tablet of paper and his own unique mind.

By 1905 he had produced no fewer than five papers, all of which were published that year by the German Yearbook of Physics, and three of which were of major importance. That same year he also earned his Ph.D.

One of the papers attacked a mysterious phenomenon, known as the photoelectric effect, that had been observed for several years: When light falls on certain metals, electrons are emitted. (Roughly similar to the principle that makes "electric eye" doors fly open when they "see" you coming.) So far, no one had been able to explain it; although the Hungarian-German physicist Philipp von Lenard [LAY nahr] found out in 1902 that there was no relationship between the intensity of the light and the energy of the electrons emitted. And a brighter light might cause more electrons to be emitted, but they would not be any more energetic than those released by a dim light. Classical physics could offer no explanation.

That's where Einstein stepped in, breaking out Planck's quantum theory, which had been gathering dust for a couple of years without too much attention. Planck had pointed out that light emits distinct "packets"; Einstein added that light also travels in packets. Einstein pointed out that a particular wavelength of light is made up of quanta of fixed energy content, according to quantum theory. When a quantum of energy bombards an atom of a metal, the atom releases an electron of fixed energy content and no other. A brighter light would contain more quanta, still always of fixed energy content, causing the emission of more electrons, also still all of the same energy content. The shorter the light's wavelength (and the higher the frequency), the more energy contained in the quanta and the more energetic the electrons released. Very long wavelengths (of lower frequency) would be made up of quanta having much smaller energy content, in some cases too small to cause any electrons to be released. And this threshold would vary depending on the metal.

This was the first use of Planck's theory since its invention to explain the blackbody problem—and once again it succeeded in explaining a physical phenomenon where classical physics could not. For this work, Einstein received the 1921 Nobel prize in physics. It was the first major step in establishing what would become known as quantum mechanics, the recognition of the discrete and discontinuous nature of all matter, especially noticeable on the scale of the very small.

In classical mechanics, energy and matter can be thought of as moving along a ramp, in continuous gradations, whereas in quantum mechanics energy and matter move in stair steps. According to quantum theory, an object can only increase or decrease its energy level by absorbing or emitting enough energy to allow it to exist at another allowed energy level. In a quantum jump, between one "stair step" and the next, matter or energy does not exist except at the allowed energy levels.

Classical mechanics always seemed to hold true as long as investigations didn't involve the very small or the very large and very fast. Planck's quantum theory helped explain how things work on the minute scales of atoms and their particles.

Einstein would become famous, however, for turning his attention to the realm of the very large—and the very fast, that is, the speed of light. But first, a word about another famous paper he published that year—on another problem that had been bothering physicists for decades.

BROWNIAN MOTION

Seventy-eight years earlier, one day in 1827, the Scottish botanist Robert Brown was observing plant pollen suspended in water under a microscope, when he noticed the irregular movement of some of the individual pollen grains. Maybe it was evidence of a "life force" within the tiny nodules. But then Brown noticed the same motion when he examined nonliving dye particles suspended in water: erratic, aimless, random movement. However, he had no explanation for it, nor did anyone for three-quarters of a century.

Now Einstein showed mathematically that molecules in constant motion in water could push minute particles enough to make them jiggle. He worked out the effect of different sizes of molecules and the angles of motion, developing an equation that could be used to work out the size of the bombarding molecules and the atoms that composed them. A few years later, 1908–09, Jean-Baptiste Perrin [peh RAN] conducted a series of experiments that served to substantiate the existence of the atom in observational terms and confirmed Einstein's theoretical work. It was the first time anyone had offered purely observational, not just deductive, evidence of the atom's existence.

THE SPECIAL THEORY

Surprisingly, Einstein never received a Nobel prize for the most important paper of the five that he published in 1905, the one that dealt with a theory that came to be known as the special theory of relativity. It's called "special" because it deals with a special case; in this theory Einstein is only talking about what happens when objects are traveling at a constant speed in a straight line (uniform unaccelerated motion). Michelson and Morley, you'll recall, could not detect any changes in the velocity of light. Einstein had wondered about this independently, and he began with the assumption that the velocity of light in a vacuum is always constant. It doesn't matter if the source moves, and it doesn't matter if the observer measuring the light is moving.

He also tossed out the idea of the ether, which Michelson and Morley had called into question. Maxwell needed it because he thought light traveled in waves, and if that were so, he thought, it needed some medium in which to travel. But what if, as Max Planck's quantum theory stated, light travels in discrete packets or quanta? Then it would act more like particles and wouldn't require any medium to travel in.

By making these assumptions—that the velocity of light is a constant, that there is no ether, that light travels in quanta and that motion is relative—he was able to show why the Michelson-Morley experiment came out as it did, without calling the validity of Maxwell's electromagnetic equations into question.

So basically, in the special theory of relativity Einstein revamped Newtonian physics such that when he worked out the formulas, the relative speed of light always stayed the same. It never changes relative to anything else, even though other things change relative to each other. Mass, space and time all vary depending upon how fast you move. As observed by others, the faster you move, the greater your mass, the less space you take up and the more slowly time passes. The more closely you approach the speed of light, the more pronounced these effects become. If you were an astronaut traveling at 90 percent of the speed of light (about 167,000 miles per second), you could travel for five years (according to your calendar-watch) and you'd return to Earth to find that 10 years had passed for the friends you'd left behind. Or, if you could rev up your engines to travel at 99.99 percent of the speed of light, after traveling for only 6 months you'd find that 50 years had sped by on Earth during your absence.

So the laws of relativity say that time is relative; it does not always flow at the same rate. For example, moving clocks slow down. In the 1960s a group of scientists at the University of Michigan took two sets of atomic clocks with an accuracy to 13 decimal places. They put one set on airplanes flying around the world. The other identical set remained behind on the ground. When the airplanes with the clocks landed, and those clocks were compared to the clocks that stayed still, the clocks that had ridden on the airplanes had actually ticked fewer times than those that had stayed on the ground.

Relativity also says that the faster an object moves, the more its size shrinks in the direction of its motion, as seen by a stationary observer. And the mass, as seen by the same observer, would seem to increase. Further, nothing, according to relativity, can ever reach the speed of light (or, more precisely, the speed at which all electromagnetic radiation travels in a vacuum—including radio waves, X rays, infrared rays and so on). The speed of light is the upper-limit ceiling because, as objects approach the speed of light, their mass approaches infinity.

Most astounding of all, using his famous equation, $E=mc^2$, Einstein also showed that energy and mass are just two facets of the same thing. In this equation, E is energy, m is mass and c^2 is the square of the speed of light, which is a constant. So the amount of energy, E, is equal to the mass of an object multiplied by the square of the speed of light.

All this seems completely against all common sense. But common sense is based on everyday experience, and things don't get really strange with relativity until you venture into the very, very fast. There, most of us have no experience to rely on. But, no matter how counterintuitive it may seem, remember that every experimental test of this theory for some 90 years has confirmed that Einstein was right.

THE GENERAL THEORY

Strangely enough, it was another four years after Einstein's publication of his papers on the photoelectric effect, Brownian motion and the special theory of relativity before he succeeded in securing a teaching position at the University of Zurich—and a poorly paying one at that. But by 1913, thanks to the influence of Planck, the Kaiser Wilhelm Institute (KWI) near Berlin created a position for him. Ever since his 1905 publications, Einstein had been working on a bigger theory: his general theory of relativity. The special theory had applied only to steady movement in a straight line. But what happened when a moving object sped up or slowed down or curved in a spiral path? The more general case of accelerated movement was much more complicated, but a theory that explained it would be even more useful.

Now, at the KWI, Einstein had the opportunity to finish the work. And in 1916, he published his general theory of relativity, which had vast implications, especially on the cosmological scale. Many physicists consider it the most elegant intellectual achievement of all time.

The general theory preserves the tenets of the special theory while adding a new way of looking at gravity—because gravity is the force that causes acceleration and deceleration and curves the paths of moons around planets, of planets around the sun, and so on.

Einstein realized that there is no way to tell the difference between the effects of gravity and the effects of acceleration. So he abandoned the idea of gravity as a force and talked about it instead as an artifact of the way we observe objects moving through space and time. According to Einstein's relativity, a fourth dimension—time—joins the three dimensions of space (height, length and width), and the four dimensions together form what is known as the space-time continuum.

To illustrate the idea that acceleration and gravity produce essentially the same effects, Einstein used the example of an elevator, with its cables broken, falling from the top floor of a building. As the elevator falls, the effect on the occupants is "weightlessness," as if they were aboard a spaceship. For that moment they are in free fall around the Earth. If the people inside couldn't see anything outside the elevator, they would have no way to tell the difference between this experience and the experience of flying aboard a spaceship in orbit.

Einstein made use of this equivalence to write equations that saw gravity not as a force, but as a curvature in space-time—much as if each great body were located on the surface of a great rubber sheet. A large object, such as a star, bends or warps space-time, much like a large ball resting on a rubber sheet would cause a depression or sagging on its surface. The distortions caused by masses in the shape of space and time result in what we call gravity. What people call the "force" of gravity is not really a characteristic of objects like stars or planets, but comes from the shape of space itself.

In fact, this curvature has been confirmed experimentally. Einstein made predictions in three areas in which his general theory was in conflict with Newton's theory of gravity:

(1) Einstein's general theory allowed for a shift in the perihelion (the point nearest the Sun) of a planet's orbit. (Such a shift in Mercury's orbit had baffled astronomers for years.)

(2) Light in an intense gravitational field should show a red shift as it fights against gravity to leave a star.

(3) Light should be deflected by a gravitational field much more than Newton predicted.

The first of these predictions was not overwhelmingly impressive, since Urbain Leverrier had already observed the shift in Mercury's orbit and in

1845 had postulated the existence of another inner planet to explain it. But Einstein's theory did solve the mystery.

As for the red shift, Walter Sydney Adams (1876–1956), an American astronomer, had by 1915 become an expert in stellar spectra. He had already shown that a star's luminosity could generally be determined from its spectrum, and he was in the midst of studying the companion star of the star Sirius. He had determined from its spectrum that this star, though very dim, was intensely hot. Such extreme heat should produce a very great luminosity, or brightness, in relation to the star's surface area, which, given how extremely dim the star is, could only mean that the companion of Sirius had a very small surface area and was enormously dense—denser than ordinary matter could possibly be. Based on Rutherford's idea that atoms are mostly empty space, though, astronomers concluded that this star might be composed of collapsed atoms, the subatomic particles of which were crushed together. British astronomer Arthur Stanley Eddington (1882–1944) suggested to Adams that this white dwarf, as this type of superdense body came

British astronomer and physicist Arthur S. Eddington was a foremost supporter of Einstein's relativity theories. He participated in one of the expeditions in 1919 to view the solar eclipse to help establish the general theory, and he suggested that W. S. Adams's white dwarf companion of Sirius was a good candidate to test the red shift predicted by the general theory. Einstein considered Eddington's 1923 treatise on relativity to be the best written in any language.
(Astronomical Society of the Pacific)

Einstein in his study at the Institute for Advanced Study in Princeton, New Jersey
(Institute for Advanced Study)

to be called, must have an extraordinarily intense gravitational field, which made it a good candidate to test the second prediction of Einstein's theory, the red shift. Sure enough, in 1925 Adams had a chance to search for this shift, and there it was. The spectral absorption lines of the white dwarf were in fact shifted more toward the red end of the spectrum than normal. More recently, in the 1960s, more sophisticated measuring devices made it possible to test the much smaller shift of our own sun's light, which was also found to corroborate Einstein's prediction.

The general theory was born in the midst of World War I, so testing for the third prediction, the gravitational deflection of light, was postponed until after the war. But in 1919, the Royal Astronomical Society of London organized two expeditions—one to northern Brazil and the other to Príncipe Island off the coast of West Africa—to take advantage of a solar eclipse that occurred at exactly a time when many more bright stars appeared near the Sun than usual. On March 29, 1919 the eclipse occurred, and, in the darkened daytime sky, the measurements of the nearby stars were taken. Then they were compared with those taken in the midnight sky six months earlier when the same stars had been nowhere near the Sun. Einstein was proved right. He rapidly became the most famous scientist in the world, and his name became a household word.

Meanwhile, Germany—long one of the premier cradles of great work in all the sciences—rapidly became less and less hospitable to the large group of outstanding scientists who worked there, especially the many who, like Einstein, were counted among the Nazis' Jewish targets. By the 1930s an exodus had begun, including many non-Jewish scientists who left on principle, no longer willing to work where their colleagues were persecuted. In 1930 Einstein left Germany for good. He came to the United States to lecture at the California Institute of Technology, and he never went back to Germany afterward. He accepted a position at the Institute for Advanced Study in Princeton, New Jersey, where he became a permanent presence, and in 1940 he became an American citizen.

Always a catalyst among his colleagues for thoughtful reflection, Einstein remained active throughout his life in the world of physics. But even this renegade found, as Planck did, that physics was changing faster than he was willing to accept. On the horizon loomed challenges to reason that he was never able to accept—such as Niels Bohr's complementarity and Werner Heisenberg's uncertainty principle. "God does not play dice with the universe," Einstein would grumble, or "God may be subtle, but He is not malicious." During the last decades of his life Einstein spent much of his time searching for a way to embrace both gravitation and electromagnetic phenomena. He never succeeded, but continued to be, to his final days, a solitary quester, putting forward his questions to nature and humanity, seeking always the ultimate beauty of truth.

THE NEW UNIVERSE, PART TWO: THE QUANTUM SURPRISE

*F*ueled with the exciting discoveries of radioactivity, quantum theory and relativity, the first 25 or 30 years of the 20th century witnessed an enormous fertility of ideas and discoveries unparalleled in the history of physics. A dynamic cluster of men and women—ambitious, brilliant, keenly prepared and talented—gathered in the universities of Europe, Britain and, to a lesser degree, Canada and the United States to ride the crest of a great wave of exploration into the inner regions of the atom. One of the finest of these minds was a young man who came to be known as the Gentle Dane.

NIELS BOHR'S ATOM

Niels Bohr (1885–1962) was a crackerjack soccer player when he entered the University of Copenhagen in 1903—although he was not as good as his brother, Harald (who made the 1908 Danish soccer team, which won second place in the Olympics). Both brothers were also brilliant, but in his student days when anyone commented on his wizardry with mathematics, Harald would remark, "I am nothing; you ought to meet my brother Niels."

Niels was a big, rangy man with a bulldog face, an unusually large head, heavy eyebrows and hairy hands. He also had an easygoing sense of humor and a charismatic way of drawing ideas out of people and stimulating discussion. By 1921, he had overseen the establishment and construction of the Institute for Theoretical Physics in Copenhagen, which he headed at the age of 36, and there, like a magnet, he drew the best young minds from all over the world.

Niels Bohr   (Institute for Advanced Study)

As one of his young students at the institute, Otto Frisch, described him, "He had a soft voice with a Danish accent, and we were not always sure whether he was speaking English or German; he spoke both with equal ease and kept switching. Here, I felt, was Socrates come to life, tossing us challenges in his gentle way, lifting each argument to a higher plane, drawing wisdom out of us which we didn't know we had, and which of course we hadn't." As a teacher and mentor, he was unrivaled. But all that was later in his career.

Earlier, as a young post-graduate, Bohr traveled to Cambridge, and then to Manchester in 1912, where he studied for four years before returning to Copenhagen as professor of physics. There, Bohr was perhaps the only theorist ever to hit it off with Ernest Rutherford. But they were a strange pair, Rutherford brash and exuberant, Bohr speaking not much above a whisper, digging in his mind for the perfect word and, in the words of C. P. Snow, "on not finding it, . . . [leaving] pauses, minutes long, in which he reiterated a word which was clinging to his mind." The contrast between the two ways of thinking became a hallmark of the first decades of the century, and no two people better typified the dichotomy than Bohr and Rutherford. Bohr the ruminator, with enormous powers of concentration, thought things through as he talked and frequently would spontaneously hit upon an idea in the middle of a conversation. For overcoming the natural resistance of apparatus to do one's bidding, however, he had no talent. Rutherford, by contrast, had the intractible persistence needed to pursue a course of action and see it through to its outcome, but he lacked Bohr's ability to daydream with purpose. In solving the "serious problems of physics," Frisch later recalled that Bohr "moved with the skill of a spider in apparently empty space, judging accurately how much weight each slender thread of argument could bear."

Bohr had not been at Manchester long when he came up with an improvement on Rutherford's 1911 model of the atom. In Rutherford's atom, electrons orbited a central nucleus, held there by its electrical attraction, much like a tiny planetary system. But the model had a fundamental problem. In the 19th century, Michael Faraday and Maxwell had shown that an electrically charged particle gives off radiation if it is diverted from a straight path. As it gives off energy, without a mechanism for gaining energy back, an electron moving in the circular orbits postulated by Rutherford, would soon spiral inward to the nucleus. That is, to satisfy the laws of conservation of energy, the orbits would have to shrink. Rutherford couldn't explain why atoms didn't just collapse. This problem didn't concern Rutherford, however. He was not a theoretician. That was where Bohr came in.

He sat up long hours, poring over experimental data, using his slide rule (this was long before computers and calculators) and jotting out equations. What if, thought Bohr, we applied Planck's quantum theory to the atomic model? In the 19th century, physicists had discovered that each element

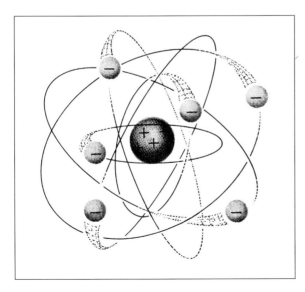

In 1913, Niels Bohr proposed a model of the atom with electrons confined to specific circular orbits around the nucleus (G. Tyler Miller, Jr., Chemistry: A Basic Introduction, 2nd Edition. Wadsworth Publishing Company, 1981.)

produces a characteristic spectrum of light when heated. That is, sodium, for instance, emits light only at particular wavelengths—yellow, in this case. Potassium emits a violet light. And so on. In terms of Planck's theorem, that meant that the atoms of each element produce light quanta only of a particular energy. Bohr worked on a model of the atom that explained why.

And he succeeded. Bohr proposed that electrons could not orbit an atom's nucleus willy-nilly in just any orbit. Because all atoms of one element

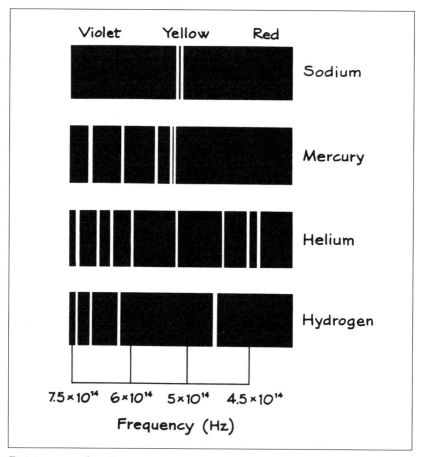

Emission spectra for sodium, mercury, helium and hydrogen. Each element displays a unique discrete spectrum, almost like a fingerprint, a fact that puzzled scientists for years until Bohr came up with his model of the atom. Bohr suggested that electrons release energy only while making jumps from one orbit to another (explaining the discrete spectral lines emitted by the elements), and that the electrons move only in certain allowed orbits at specific distances from the nucleus of each element. (Jacqueline D. Spears and Dean Zollman, *The Fascination of Physics.* Benjamin/Cummings, 1985.)

function alike and therefore doubtless look alike he suggested that electrons of any element move only in certain allowed orbits at specific distances from the nucleus. The radius of the orbit was determined by Planck's constant—and therefore, so was the energy. As long as electrons move within the allowed orbits, he said, they do not emit electromagnetic energy. But electrons could jump spontaneously from one permissible orbit to another, and then they would change energy states, absorbing or releasing energy in packets, or quanta. Moving inward, toward the nucleus, into an orbit having a smaller radius, they would give off energy. Moving out in the other direction, they would absorb energy.

Bohr did the calculations for hydrogen's single electron. He worked out the energies involved for jumping from one orbit that was permissible to another. He calculated the light wavelengths that would be produced if the energy were converted to light (photons, or quanta of electromagnetic energy). It worked. His calculations matched the spectrum of hydrogen, which had always been a mystery before. Physicists had observed that certain spectra were associated with the atoms of certain elements, but they had never before been able to explain why. Bohr did so in precise detail.

It was a great coup. When Einstein heard about how well the data matched with the spectra, he exclaimed, delighted, "Then this is one of the greatest discoveries." Bohr became the father of atomic theory.

But even Bohr, having put quantum to its first use in the physics of matter, recognized the theorem's enormous mysteries.

"It makes me quite giddy to think about these problems," a visitor once complained to Bohr. "But, but, but . . . ," Bohr stammered ingenuously, "if anybody says he can think about quantum theory *without* getting giddy it merely shows that he hasn't understood the first thing about it!"

Of course, Bohr's model of the atom is by no means the last word. Our ideas about the atom have changed a great deal since his announcement in 1913, and among those who contributed to their development was an eccentric and precocious young man from Germany.

PAULI'S EXCLUSION PRINCIPLE

Wolfgang Pauli was clumsy in the lab and faltering in the lecture hall. He was short and pudgy, and somewhat resembled the movie actor Peter Lorre. But his mind could pierce to the heart of a problem practically without effort. He studied under Arnold Sommerfeld at the University of Munich, where he did his doctoral work, and then pursued postgraduate studies both in Copenhagen with Bohr and at Göttingen. Later he moved to the United States, joining the Institute for Advanced Study in Princeton, and he became a U.S. citizen in 1946.

Wolfgang Pauli
(Photo by S. A. Goudsmit,
Courtesy AIP Emilio
Segrè Visual Archives)

His great insight, known as Pauli's exclusion principle, came to him as he was battling a problem known as "the Zeeman effect" (named after Dutch physicist Pieter Zeeman). He was not immune to frustration as a scientist: Moody and dejected while visiting Bohr, he once snappishly replied to Margrethe Bohr's solicitous inquiry, "Of course I am unhappy! I cannot understand the anomalous Zeeman effect."

Pauli based his work on an enormous pile of data, in which he discerned a simple sorting-out principle that held true in all cases: In any system of elementary particles—for example, the collection of electrons within the atom—no two particles may move in the same way, that is, occupy the same energy state. He announced his exclusion principle in 1925, and it has since held true for dozens of other nuclear particles that, at the time, no one had even dreamed of yet. The concept became an important part of the future of quantum mechanics.

The exclusion principle explained why not all the electrons in an atom drop down to the orbit nearest the nucleus, where the least amount of energy is required to complete an orbit. Once an electron is in an orbit, it excludes any other electron from occupying the same orbit. Pauli received the Nobel prize for physics for this work in 1945 (somewhat belatedly).

Pauli also solved another mystery: When beta particles (speeding electrons) were emitted by atoms, some energy seemed to be missing. This situation clearly violated the law of the conservation of energy, and physicists were reluctant to let go of a principle that worked so consistently everywhere else. In 1931 Pauli postulated that another, tiny particle without charge and possibly without any mass was emitted at the same time as beta particles and carried off the energy that seemed to be "missing." The following year Enrico Fermi named this particle a *neutrino*, which is Italian for "little neutral one." Neutrinos are so small that they were almost impossible to detect with existing technology, and for years no one could prove their existence. There were those who suspected that Pauli had just pulled a sort of bookkeeping trick—had invented a particle to make the energy books look balanced. But in 1956, through the use of a nuclear power station, an elaborate experiment was completed that proved the existence of the ghostly neutrino, and Pauli was vindicated.

PARTICLE AND WAVE

A coterie of young, very bright physicists seemed to have the world by the tail in 1925, the year Pauli came out with his exclusion principle. In Paris two years earlier, Louis de Broglie [BROH glee] had proposed that if subatomic particles were also—at the same time—thought of as waves, the consequences theoretically worked out very neatly. It was a simple but novel idea, the kind that leaves you saying, "Well, what *about* that? What if that *were* so?" According to Planck and Einstein, light, which had most recently been regarded as a wave, should be regarded as a particle. Now de Broglie was saying that particles—electrons, even atoms—could sometimes behave like waves. When his theory was tested experimentally, it became clear that he was right. This mind-boggling concept is referred to as the wave-particle duality.

The idea immediately caught on among physicists. Erwin Schrödinger worked out the mathematics of De Broglie waves. It worked. This was another way of looking at the atom. At first confusion reigned. Which was right, quanta or wave/particles? Finally Schrödinger showed that the two formulations were equivalent mathematically, publishing his work in 1926. While this may not have been a satisfying explanation to everyone, it made the physicists happy. It was a mathematically sound atomic theory.

There was only one thing wrong: Schrödinger had come up with the idea that electrons *were* waves—sort of "matter waves." And his equations had worked perfectly. But something didn't quite fit. Then, that same year, another German physicist, Max Born, came up with the idea that what Schrödinger had described with his equation was, not the electron itself, but the *probability* of finding the electron in any given location. For example, if you bombard a barrier with electrons, some will go through and some will bounce off. (You've seen this phenomenon with light on a window. Some

Erwin Schrödinger
(Photo by Francis Simon,
Courtesy AIP Emilio
Segrè Visual Archives)

of it bounces off, and you see a reflection; some of it penetrates, and you see through the window, as well.) Born saw that you could figure the odds—a single electron has, say, a 55 percent chance of going through the barrier, and a 45 percent chance of bouncing back. Because electrons cannot readily divide, Schrödinger's wave equation couldn't have been describing the electron itself—only its probable location.

Leon Lederman, one of the world's current top physicists, calls Max Born's interpretation "the single most dramatic and major change in our world view since Newton." Schrödinger wasn't happy with it at all, and neither were a lot of other classical physicists of his day. Born's probability meant that the determinism promised by Newton's laws of physics was now out the door. Coupled with quantum theory, it meant that only probabilities could be known—about anything you wanted to measure.

But Bohr, Sommerfeld, Heisenberg and others took Born's ideas in stride—the concepts seemed to fit—and they and their colleagues continued the exciting work of trying to get *all* the pieces to fit. Among those engaged in the challenge was English physicist Paul Dirac, still in his twenties, who came up with an elegant new equation for the electron, based on his efforts to get quantum and relativity to work together. (It came to be called the Dirac equation.) When he found the solution to his equations in 1930, they also showed the astonishing conclusion that wherever matter existed, its mirror must also exist, which he called antimatter. For example, another particle had to exist with the same properties as the electron, with one important exception: Instead of the electron's negative charge, this particle should have a positive charge. His line of reasoning was related to Emmy Noether's ideas about symmetry (see box) and the fact that the square root of a number can be either a plus or minus number (the square root of 4, for example, is both $+2$ and -2; $2 \times 2 = 4$ and $-2 \times -2 = 4$). Look, Dirac's equations said, for a positive electron. Then, in 1932, a young physicist named Carl Anderson was working with a powerful magnet and a cloud chamber at California Institute of Technology, when he saw it—at least he saw the tracks of a subatomic particle that seemed to be an electron, but was pulled the opposite way by the magnet. He called his new particle a "positron."

Werner Heisenberg
(Photo by Friedrich Hund, Courtesy AIP Emilio Segrè Visual Archives)

THE ROLE OF UNCERTAINTY

Meanwhile, in 1927, Werner Heisenberg produced another amazing physical theory: the uncertainty principle. By that he meant that the exact position and precise velocity of an electron could not be determined at the same time. Or, you can't tell for certain where an electron will go when you

AMALIE EMMY NOETHER: SYMMETRY IN MODERN PHYSICS

More and more in the 20th century mathematics has come to play a key role in the formulation of theory in physics. A brilliant mathematician named Amalie Emmy Noether was the first to establish the role of symmetry (or mirror reflections) in modern physics. For her achievements she is recognized as one of the most important mathematicians of this century.

Noether was born in 1882 in Germany in the university town of Erlangen. Despite university rules that prohibited women from entering the university, Noether managed to gain permission to attend lectures. Only one other woman attended lectures there at the time—out of 968 students—and neither of them was permitted to enroll. Finally, in 1904, the regulation was lifted, and Noether received her degree; in 1908 her doctoral thesis was accepted *summa cum laude* (with highest honors).

Noether loved abstract mathematics, and that was her life. She taught for several years in Erlangen without pay and often substituted as lecturer for her father, who was a mathematics professor at the university. In 1915, she worked at Göttingen on the mathematical equations for Einstein's general theory of relativity, still without pay. A faculty member there, mathematician David Hilbert, sometimes let her teach in his place, and he championed her for a (still unpaid) position with the arch-conservative faculty, declaring, "I do not see that the sex of the candidate is an argument against her admission as *Privatdozent* [teaching assistant]. After all, we are a university and not a bathing establishment." Inclined to forget about her appearance, she usually looked rumpled, disheveled and overweight— a sort of female Einstein, with thick glasses and a love of argument.

The important part she played in the development of particle physics was only a fraction of the total volume of her productive career, and she did some of her best work in later life—an unusual feat for both mathematicians and physicists. She had a rare facility for looking beyond the surface of problems to the bedrock on which she could base her mathematical interpretations.

hit it—you can only say where it probably will go. One could only make statistical predictions.

This idea capped off the great scientific revolution we call quantum theory. Much remained to be resolved, however, and quantum field theory is still evolving today. Some scientists contend that the theory will not be complete until it is fully combined with gravitation.

The theory she contributed to physics came as an offshoot of David Hilbert's interest in relativity and the early work she did on that subject in 1915. It demonstrates that symmetry in nature and conservation laws go hand-in-hand. Wherever you find one, you'll also find the other.

A symmetry is like a reflection in a mirror—in fact a mirror image *is* a kind of symmetry. In mathematical terms, the coordinates of the original have been changed from z to -z, assuming that z is the axis pointing toward the mirror. What Noether pointed out is that something always remains unchanged in symmetry. In the mirror image example, the x and the y coordinates remain the same. And this sameness (invariance) is what we mean by conservation. For example, the symmetries of space and time are linked with conservation of energy, momentum and angular momentum. But that's not all. Each also *implies* the other. Conservation laws necessarily result from symmetries, and symmetries necessarily entail conservation laws.

Many physicists have profited from these ideas. On the simplest level, Noether's work on symmetry provided what proved to be a reliable roadmap. Across from this, look for that. Her ideas on symmetry and invariance proved very useful to Albert Einstein in his work on relativity and the importance of Noether's work has continued to gain recognition as physicists have discovered more than a dozen conservation laws and symmetries associated with them. It has become one of the foundations of modern physics.

Like many other Jewish intellectuals, Noether fled Germany in 1933 after the Nazis came to power, and she left behind her teaching position at the University of Frankfurt. In the United States, she became a visiting professor at Bryn Mawr College, a women's school outside Philadelphia, and lectured at the Institute for Advanced Study in Princeton.

In 1935, Noether died unexpectedly after surgery. Her friend, mathematician/physicist Hermann Weyl, delivered the memorial address, in which he remarked, "She was not clay, pressed by the artistic hands of God into a harmonious form, but rather a chunk of human primary rock into which he had blown his creative breath of life."

Einstein never accepted the uncertainty principle and he debated it long and ardently with Bohr. It was an issue that the two friends, always respectful of each other's intellect, pursued to the end of Einstein's days. Years after Einstein died, Bohr was still fashioning illustrations to convince him. And the day Bohr died his blackboard still showed a diagram he had put there the night before in his mental discussion with his old friend.

In fact, Bohr proposed his own contribution to the discussion, known as complementarity, in 1927, maintaining that a phenomenon can be looked upon in two mutually exclusive ways. Yet both outlooks can remain valid in their own terms. As his student Frisch put it, "It is a bit as if reality was painted on both sides of a canvas so that you could only see one aspect of it clearly at any time." Einstein was never comfortable with this concept either.

Brilliant, vital and exciting ideas, they are nonetheless unsettling, to be sure. Of quantum, the theoretical physicist Richard Feynman use to say to his students, "I think I can safely say that no one understands quantum mechanics. . . . Do not keep saying to yourself, if you possibly can avoid it, 'But how can it be like that?' because you will get 'down the drain,' into a blind alley from which nobody yet has escaped. Nobody knows how it can be like that."

These, of course, are not the only names that built the atomic and quantum edifice during these exciting years—they are only some of the more prominent ones. Many more contributed, by counting thousands of scintillations as electrons bumped against a screen, devising apparatus, offering ideas, stimulating new perspectives. More and more, science was becoming not the individual effort of a Copernicus in his tower or a Tycho Brahe alone in his observatory, but teamwork. Many of the heroes and heroines remain unknown, and of those who are known, there are too many to fit into one short book. But teamwork—experimentalists verifying theoreticians and theoreticians bouncing ideas off the hard data of research—had become more and more central to the way science was done.

But one area where the lone scientist still prevailed in many ways was the study of the skies and the cosmos.

CHAPTER 4

NEW OBSERVATIONS
OF THE UNIVERSE

*I*t was March 10, 1910, one of those cold, blustery days in Paris, and it was frigid at the top of the Eiffel Tower. Built only 21 years before, the huge steel girders stretched nearly 1,000 feet up into the sky, by far the tallest structure on the Paris horizon. On this particular day, a Jesuit physics teacher from Valkenburg in the Netherlands emerged from the elevator pulling his equipment out onto the observation platform. He was not the usual sightseer. High above the Champ de Mars, Father Theodor Wulf used his glass and metal instruments to determine how well the air at that altitude conducted electricity.

What he found was a surprise—since air does not usually conduct electricity at all. But Wulf was one of the "radioactive people," a term applied to those who studied the fascinating radiation discovered by Becquerel in 1896. Using an electroscope, which looked something like an antenna in a jar, one could measure the strength of a radioactive source. Held near uranium, its strips of metal foil would spread apart and fall back together as they discharged into the air around them. The faster they discharged, the stronger the source. But, as Wulf found that day in March, sometimes these instruments seemed to "leak," slowly discharging even when no uranium lump was nearby. This residual discharge confused readings, but no one could find a way to get rid of it. In 1909, Wulf had invented a highly sensitive electroscope, which of course was even more likely to show a residual discharge because of its precision. What was the source of this strange phenomenon? Geologists, meteorologists and physicists all over the world began testing with Wulf electroscopes. Wulf tested locations in Germany, Austria and high in the Swiss Alps. Residual discharge seemed to be everywhere, but in different amounts. Was it radioactivity escaping from the crust of the Earth? This was what Wulf climbed the Eiffel Tower to test. From high on this steel structure, with 1,000 feet between his instrument and the Earth's crust, he should be able to eliminate any radioactive reading coming from the Earth itself. He spent four days testing. The electroscope never stopped discharging in all that time. There must be, he concluded, "either another source [of radioactive emissions] in the upper portions of the atmosphere, or their absorption by the air is substantially weaker than has hitherto been assumed."

At about this time, Victor Hess, from the newly founded Vienna Institute for Radium Research, joined the fray. Between 1911 and 1913, he took his electroscope up in balloons, high into the atmosphere, no fewer than 10 times. Typically the discharge slowed as he ascended, then leveled off. It did not stop or drop as much as it should have if the source of radiation were the Earth's crust. The puzzle continued. Then, on the ninth flight, Hess made an extraordinary measurement. At 15,000 feet, he noticed that the rate of discharge had doubled the rate that was common on the ground. He came to a strange, even eerie conclusion: "that rays of very great penetrating power are entering our atmosphere from above," from deepest space.

No one believed this outlandish idea at first. Then on June 28, 1914, a German named Werner Kolhörster made an ascent to a record 30,000 feet—higher than Mt. Everest. The ionization level there was 12 times that at sea level. That same day, however, World War I broke out and the testing came to a halt. But the confirmation had been made. Hess was right: Powerful radiation was continually bombarding our planet, and everything else in the universe. Cosmic rays had finally been discovered. It was a new insight that boggled the mind.

In astronomy and astrophysics, as in physics, scientists were developing new tools and finding new ways to collect data, measure and interpret. They began to increase the use of photography and stellar spectra to gather data, devise new systems for classification and interpretation, and develop insights into the rapidly multiplying complexity of observations they made.

UNDERSTANDING THE UNIVERSE

Karl Schwarzchild [SHVAHRTS shild] (1874–1916), who became a professor at Göttingen in 1901, became one of the pioneers in the use of photography for measuring the brightness of stars, particularly of variable ones. He also suggested that periodic variable stars (those that vary their brightness, or luminosity, on a periodic cycle) behaved the way they do because of temperature changes that occurred periodically.

Schwarzchild, like everyone else interested in understanding the universe, was intrigued by Einstein's theories, and he was the first to offer a solution

Ejnar Hertzsprung   (Photo by Dorritt Hoffleit, Yale University Observatory, Courtesy AIP Niels Bohr Library)

to Einstein's field equations. He was also the first to calculate gravitational phenomena in the neighborhood of a star with all its mass concentrated in a point—an arrangement that came to be called a *black hole*. Schwarzchild's estimation of the boundary of a black hole, called the Schwarzchild radius, is still accepted.

Making Sense of Stars

Devoted to the idea of lecturing and writing on popular astronomy, Schwarzchild was responsible for bringing Ejnar Hertzsprung [HURT sprung], an amateur astronomer and popularizer, to Göttingen as a professor of astrophysics in 1909. Hertzsprung, who was educated as a chemical engineer, worked for two years in St. Petersburg, then returned to Copenhagen in his native Denmark in 1902, where he did considerable work as an amateur astronomer. Hertzsprung was bothered by the fact that a nearby dim star can appear to be brighter than a faraway bright star; to compensate for this, he came up with an idea he called "absolute magnitude," referring to the intrinsic luminosity of a star—not its apparent brightness as seen by the observer. He invented a system for comparing the brightness of stars by imagining them all at the same distance from the observer—a distance of 10 parsecs, or a little over 190 trillion miles. (A parsec is the distance at which a star has a parallax of one second.)

He also studied the relationship of color and luminosity among stars as early as 1905. Hertzsprung was a specialist in astrophotography, and he had worked at estimating stellar magnitude from photographs and photographing double stars accurately. He published his work semi-popularly, however, so it was overlooked for several years by the academic community. Henry Norris Russell (1877–1957), an American astronomer, announced his similar results, found independently, in a more formal way, and the two usually share the credit for what is now known as the Hertzsprung-Russell diagram of star luminosity. The purpose of the diagram is to arrange and study the data about star forms in order to track their relationships, and it remains an important tool in understanding the types of stars and comparing them objectively, based on physical variables.

Hertzsprung discovered in 1911 that the Pole star is a Cepheid variable and in 1913 he was the first to estimate the actual distances of some Cepheid variables, a certain type of pulsating star. This, together with the work of Henrietta Leavitt, allowed Harlow Shapley to work out the proper shape of our galaxy.

Looking Inside Stars

One of the great mysteries that astronomers sought to solve was how we can determine the internal structures of stars. What's happening inside them? And why are there different types? A. S. Eddington commented in 1926:

> At first sight it would seem that the deep interior of the sun and stars is less accessible to scientific investigation than any other region of the universe. . . . What appliance can pierce through the outer layers of a star and test the configurations within?

> The problem does not appear so hopeless when misleading metaphor is discarded. It is not our task to "probe"; we learn what we do learn by awaiting and interpreting the messages dispatched to us by the objects of nature. And the interior of a star is not wholly cut off from such communication. A gravitational field emanates from it. . . . Radiant energy from the hot interior after many deflections and transformations manages to struggle to the surface and begin its journey across space. From these two clues alone a chain of deduction can start which is perhaps the most trustworthy because it [employs] only the most universal rules of nature—the conservation of energy and momentum, the laws of chance and averages, the second law of thermodynamics, the fundamental properties of the atom, and so on.

So physics and astrophysics moved hand in hand. Using the advances gained from the new theories in physics, Eddington was able to show why stars are the way they are. Gravity pulls inward, he said, on the stellar gas while gas pressure and radiation pressure push outward. In a stable star, he recognized, these forces are balanced.

MEASURING THE UNIVERSE

For centuries astronomers had been looking for good ways to measure the size of the universe, and with Cepheid variables, Henrietta Swann Leavitt (1868–1921) discovered a good yardstick in 1912 at the Harvard Observatory.

The first Cepheid variable was discovered in 1784 by a 19-year-old amateur astronomer named John Goodricke. Cepheids are stars that have regular variations in brightness, over regular periods usually lasting five to 30 days. These variations are as regular as clockwork, so they are more readily understood than stars with irregular variations. But more important, working with the stars in the Small Magellanic Cloud, Leavitt was able to demonstrate a remarkable relationship between the average luminosities and the periods of Cepheid variables. The period-luminosity relationship enables astronomers to figure out the luminosity of any Cepheid at any distance just by measuring its period. So, Leavitt recognized, it was easy to use this fact to measure the distances to other stars as well. Find a Cepheid and measure its period to obtain its luminosity or absolute magnitude. Then measure its apparent magnitude (how bright it appears to be) and derive its distance (and the nearby star's distance). It was an important breakthrough.

Henrietta Leavitt (Harvard College Observatory, Courtesy AIP Niels Bohr Library)

Harlow Shapley
(Photo by Frank Hogg,
Courtesy of Helen Sawyer
Hogg and Owen
Gingerich)

The Shape of the Home Galaxy

Harlow Shapley was born in Nashville, Missouri, in 1885, the son of a farmer. It's easy to imagine him as a boy, looking up at the dark Missouri skies and watching the stars. He began his career as a reporter, but he saved enough money to enter the University of Missouri in 1903, where he studied mathematics and astronomy, graduating in 1910. He continued at Princeton University, where he worked with H. N. Russell and earned his doctorate in 1913. The following year he joined the staff at the Mount Wilson Observatory in California. There he used the big, 100-inch telescope to study the globular clusters (dense, sphere-shaped clusters of stars, usually old) and began the theoretical and observational work on stellar clusters and variable stars that would make him famous. Using the relationship that Leavitt had found between apparent magnitudes and the periods of Cepheid variables, he calibrated the period-luminosity relationship in terms of

54

absolute magnitude (how bright a star would look if it were a standard distance of 10 parsecs). This became a new yardstick to determine the scale and geometry of the galaxy.

He discovered that the Sun was not at the center of the Milky Way Galaxy, as everyone had assumed, but about 50,000 light years off center. Like Copernicus, who said that the Sun, not the Earth, was at the center of the Solar System, Shapley had once again moved humankind and its home away from the center of things. His measurements also showed that the universe was much, much larger than anyone had ever thought.

After eight highly fruitful years in California, Shapley became director of the Harvard Observatory. For 31 years he oversaw the astronomy program there, expanding the staff and observation facilities and establishing a world-class graduate program that became the best in the United States.

Willem de Sitter
(Courtesy Yerkes
Observatory)

De Sitter's Expanding Universe

Willem de Sitter was a Dutch astronomer whose early interest in Einstein's theory of relativity helped popularize it. He was an influential professor of astronomy at the University of Leiden, and his reports to A. S. Eddington in England led to increased popularity for relativity there and helped encourage the expedition launched by the English to test the general theory's prediction during the solar eclipse of 1919.

De Sitter added a couple of important insights to Einstein's vision of the universe. Because light was bent by gravitational forces, he said, any ray of light would eventually curve and curve and curve, finally curving back to its starting point. The universe, de Sitter maintained, consisted of "curved space." Einstein saw the universe as curved space, but static. But de Sitter interpreted the theory differently. He saw the curvature constantly growing less and the curved universe constantly expanding outward. The spectra of the distant galaxies, which the American astronomer Edwin Powell Hubble had interpreted, confirmed this, and de Sitter won Einstein over.

Edwin P. Hubble
(Mount Wilson and Las
Campanas Observatories)

CLASSIFYING STELLAR SPECTRA:
ANNIE JUMP CANNON

Annie Jump Cannon   (Harvard College Observatory and Whitin Observatory, Wellesley College)

The daughter of a state senator, Annie Jump Cannon (1863–1941) attended Wellesley College, from which she graduated in 1884. Ten years later, she returned to school at both Wellesley and Radcliffe to do further studies in astronomy, after which she joined the staff of the Harvard Observatory in 1896. There she spent the rest of her career.

Under the direction of E. C. Pickering, the Harvard Observatory had begun an extensive study of stellar spectra using a technique he originated. Instead of trying to focus stars one at a time through a small prism, Pickering conceived of the idea of placing a large prism in front of a photographic plate. Thus each star in the field appeared on the photograph as a tiny spectrum instead of as a point of light. In this way huge amounts of data could be collected and analyzed statistically.

Cannon developed a classification system for these photographs of stellar spectra that remained in use by Harvard for more than three-quarters of a century. She found that most of the spectra could be arranged into a continuous series, identifying stars on the basis of temperature, from hottest to coolest. Her work became the basis for the *Henry Draper Catalogue,* which grew to include classifications of 225,300 stars brighter than the ninth or tenth magnitude.

Hubble's Better Yardstick

Like many astronomers of his time, Hubble (1889–1953) didn't start out to make astronomy his life's work. A Rhodes scholar at Oxford, he graduated in 1910 with a degree in law. In 1929, though, he recognized, in the process of examining nebulas and classifying galaxies, that the speed at which a galaxy recedes from us is directly proportional to its distance. Using this measurement, Hubble was able to estimate the radius of the knowable universe (that portion we can study) at 13 billion light years. That makes the diameter 26 billion light years. It was a great yardstick that worked even better than Leavitt's Cepheid variables.

THE ATOM SPLIT ASUNDER: SCIENCE AND THE BOMB

While astronomers continued their exploration of the vast reaches of the universe, physicists continued to probe the minute realm of the atomic nucleus. Austrian physicist Lise Meitner had spent a cold, lonely winter in Stockholm in 1938, having fled to Sweden's northern safety from the frightening Nazi persecution of Jewish citizens. For her, the flight from her beloved second home in Berlin had torn her work—and hence her life—apart. For 30 years she had worked side by side with the respected German chemist Otto Hahn. He typically did the experimental work, she the theoretical interpretation. Now, at age 60, her laboratory and her partner remained in Berlin, and she had almost no equipment to work with at the not-yet-built laboratory at the new Physical Institute in Stockholm, where she had been lucky even to find a position. So when Christmas holidays approached, she gladly accepted the invitation of friends to visit them on the northeast coast of Sweden in Kungälv, a little resort town where they lived. She wrote to her partner Hahn that she'd be there if he needed to reach her by the slow mails—the only way they had to communicate now. And she looked forward to seeing her favorite nephew, Otto Frisch, also a physicist, who planned to travel up from Copenhagen to spend vacation with her.

But when Meitner arrived in Kungälv, she was met with an extraordinary surprise. A letter from Hahn had preceded her there. "When I came out of my hotel room after my first night in Kungälv," Frisch later wrote, "I found Lise Meitner studying a letter from Hahn and obviously worried about it."

The two left the hotel talking excitedly, trampling through the snow in the nearby woods—Frisch on skis, Meitner at his side on foot, insisting that she could walk just as fast without skis.

Hahn and his assistant, Fritz Strassmann, had been trying to solve a mystery by bombarding small quantities of uranium with neutrons. This process typically produced only a few thousand atoms of "daughter substances," new substances having a different atomic makeup from the parent, which in this case was uranium. Then the challenge was to identify the new substances and explain why they were produced.

Hahn had begun this experiment because in 1934 Enrico Fermi, in Italy, had tried using neutrons slowed by paraffin to bombard elements with heavy uranium nuclei. Not trying to split the atoms, he hoped that some neutrons would stick to the nuclei and create isotopes with an unusual number of neutrons. What happened surprised and puzzled him, as well as everyone else. His experiment produced very large quantities of radiation. He thought he may have created a synthetic element, heavier than uranium.

No one at the time thought that it was possible to split a uranium atom. Splitting the nucleus of a lighter element was one thing—the particles were fewer and presumably less tightly bound. Splitting them didn't require a lot of energy. But no one in the early 1930s imagined it would be possible to release the tremendous energy that must bind the uranium nucleus, the heaviest natural element of them all. On this Rutherford, Einstein and Bohr were in complete agreement.

But four years later, by the time Meitner and Frisch went for their walk in the snow, strange results from various laboratories had begun to pile up. Already, in Paris, Irène Joliot-Curie (Marie and Pierre Curie's daughter) and her associate, Pavel Savitch, had produced some surprising results in this way. In a paper they had published just a few months earlier, they had asserted that while bombarding uranium with neutrons, they had produced a substance that appeared to be lanthanum—an element whose place in the periodic table was 35 steps below uranium's! That is, the element lanthanum has an atomic number (number of protons, or positively charged particles) of 57, while uranium's is 92. No one thought it was possible to split that big a chunk out of the uranium atom.

The news was highly controversial and Hahn had set out to prove that Joliot-Curie and Savitch "were very muddled up," as a visiting radiochemist heard him remark. And Hahn and Strassmann set about bombarding their own samples of uranium. But after many repetitions, they had reported their own set of surprising results in November 1938. They had found three isotopes that no one had ever identified before, which they believed to be forms of radium, element 88. This, they wrote in a November 1938 publication, "must be due to the emission of two successive alpha particles." (An alpha particle is a positively charged subatomic particle made of two protons and two neutrons.) Even though radium is not nearly as far down the table as lanthanum, Hahn and Strassmann's report, like Joliot-Curie and Savitch's, was also greeted with both skepticism and interest. Niels Bohr,

Hahn later wrote, "was skeptical and asked me if it was not highly improbable." To which Hahn heartily agreed. The whole situation was puzzling and mysterious. Was something amiss in the experiments? Hahn and Strassmann went back to their lab. The story had been emerging from his letters. Before Meitner left for vacation she had already received a letter from Hahn, written on December 19:

> As much as I can through all of this I am working, and Strassmann is working untiringly, on the uranium activities. . . . It's almost 11 at night . . . The fact is, there's something so strange about the "radium isotopes" that for the time being we are mentioning it only to you. . . . Our radium isotopes act like barium.

This was even more mystifying than the previous results. Barium has an atomic number of 56! Just a little more than half of uranium's 92.

"Perhaps you can suggest some fantastic explanation," Hahn continued in his letter. "We understand that it really *can't* break up into barium. . . . so try to think of some other possibility."

So Meitner was primed when she had found another letter from Hahn waiting at Kungälv. Further tests brought further proof that it *was* barium, Hahn wrote. As a chemist, Hahn felt he stood on firm ground. By bombarding uranium with slow neutrons, he had produced not radium, but barium. Could she come up with an alternative, more reasonable explanation? This was the problem that so perplexed Meitner and young Frisch that day in the winter snow. As Frisch later wrote:

> No larger fragments than protons or helium nuclei (alpha particles) had ever been chipped away from nuclei, and to chip off a large number not nearly enough energy was available. Nor was it possible that the uranium nucleus could have been cleaved right across. A nucleus was not a brittle solid that can be cleaved or broken . . .

In fact, the latest theory about the nucleus, set forth by the Russian physicist George Gamow and amplified by Frisch's mentor, Niels Bohr, likened the nucleus of an atom to a drop of water. During their walk in the snow, Meitner and Frisch began to postulate that a nucleus might divide into two smaller "drops" in much the same way that a drop of water might as it hung from the eaves or from the edge of an umbrella. It might slowly elongate, forming a narrow neck, and then finally separate, rather than being broken in two.

The two sat on a log, scribbling formulae on scraps of paper from their pockets. They discovered that the charge of a uranium nucleus was in fact large enough to offset the force of surface tension, making the uranium nucleus very unstable. Bombardment by a single neutron could set it off dividing. "But there was another problem," according to Frisch:

Lise Meitner in about 1959, 20 years after her famous walk in the snow with her nephew, Otto Frisch (AIP Emilio Segrè Visual Archives)

After separation, the two drops would be driven apart by their mutual electric repulsion and would acquire high speed and hence a very large energy, about 200 MeV [million electron volts, usually pronounced "emmeevees"] in all; where could that energy come from? Fortunately Lise Meitner remembered the empirical formula for computing the masses of nuclei and worked out that the two nuclei formed by the division of a uranium nucleus together would be lighter than the original uranium nucleus by about one-fifth the mass of a proton. Now whenever mass disappears energy is created, according to Einstein's formula $E=mc^2$, and one-fifth of a proton mass was just equivalent to 200 MeV. So here was the source for that energy; it all fitted!

Meitner and Frisch looked at each other in the glistening snow. While 200 MeV is not a lot of energy in everyday terms, it is a huge amount of energy for a single atom to give off. Most chemical reactions produce only about 5 electron volts. Here was a process that produced 40 million times that.

Given Hahn's results and the formulas, it became clear that Otto Hahn and Fritz Strassmann had done the undoable. They had split the uranium atom. Both Meitner and Frisch realized this as they sat in the snow in Sweden. They also realized the implications of this capability in the hands of Nazi Germany. The relatively enormous quantity of energy released by splitting atoms could have great destructive power. The conclusions Meitner and Frisch had reached were very exciting news for the scientific community, news that must reach the right people.

While Meitner returned to her lab in Stockholm, Frisch hurried back to Denmark, where he had been working in Niels Bohr's institute. Arriving just before Bohr's departure by steamer for a meeting in the United States, he set the evidence before his mentor.

"Oh what idiots we all have been! Oh but this is wonderful! This is just as it must be!" exclaimed Bohr, striking his hand to his head. Excited, he encouraged his protégé to publish a paper with Meitner on their interpretation of the Hahn-Strassmann results as soon as possible. And he embarked for America. En route, he mentioned the exciting news to a colleague, and at a meeting on January 16, 1939 the news leaked out—before Meitner and Frisch's paper was published, as it turned out (a slip-up that kind-hearted Bohr always regretted). Literally overnight, physicists and chemists in universities all over the United States began testing the premise and found it was true. The atom had been split! The worlds of physics and chemistry had been flipped upside-down.

And, while it was German scientists who first did the deed, the news had emerged into the international community of science, and it was the American scientists who would put the split atom to its first, most formidable and daunting task.

RACE AGAINST HITLER

With the news out from Otto Hahn, via Meitner, that German scientists had achieved fission, consternation mounted in the United States about the possibility that Adolf Hitler might succeed in developing an atomic bomb—the destructive potential of which would clearly be devastating. For such a weapon to fall into the hands of such an unprincipled leader in the midst of a war was unthinkable. It became more urgent than ever to beat Hitler to the draw and win the war.

In 1939 Leo Szilard [ZEE lahr] succeeded in persuading Einstein, as the most influential scientist in the world, to convince U.S. president Franklin D. Roosevelt that the United States urgently needed to begin a crash program to develop a fission bomb. This Einstein did, much against his privately held beliefs, for Einstein had always been a dedicated pacifist. But

Hitler and the Nazis had become the single most heinous force ever to gain power in the world. And so, with Einstein's reluctant backing, the Manhattan Project was born.

FERMI'S NUCLEAR REACTOR

The uranium nucleus could be split. Great amounts of energy—for an atom—could be released. But much more energy was needed for a strategic atomic weapon to work. A chain reaction would have to take place, and no one had ever set off a nuclear chain reaction before. So the first step, though no one doubted that a bomb could be built, was to demonstrate that a chain reaction could, in fact, take place.

That job fell to the talented physicist from Italy, Enrico Fermi, one of the wizards in nuclear physics, who had received the Nobel prize in physics in 1938—just months before Meitner and Frisch interpreted the Hahn-Strassmann results as fission. Fermi had taken advantage of his trip to Stockholm for the Nobel prize ceremonies to gather up his family and flee Mussolini's fascist rule. They left Italy with a few belongings and never went back, settling in the United States.

Enrico Fermi
(University of Chicago
Library, Special
Collections)

Now, beneath a set of bleachers at an athletic field at the University of Chicago, Fermi (officially classified as an "enemy alien" because he was an Italian citizen) led a team of scientists in building a test "pile," as he called it. And he put to work what he had already learned in the previous half-dozen years about the fine art of bombarding the nuclei of uranium atoms with James Chadwick's neutrons (see box on page 66).

In fact, back in Italy before the war, Fermi had pioneered the practice of using neutron "bullets" to explore atomic structure. As soon as James Chadwick announced his discovery, Fermi recognized the superiority of the neutron for his work. Alpha particles and protons have a positive charge and are repelled by the positive charge of the atomic nucleus. But neutrons have no charge—so they didn't have to be accelerated and they were easily absorbed by the nucleus. Fermi had found that slow neutrons worked best,

Enrico Fermi and a team of scientists (of which Henry W. Newson, pictured, was a member) built this and a series of some 28 other piles of uranium and graphite under the West Stands of Stagg Field at the University of Chicago in 1942. These structures of alternating "dead" layers and uranium fuel layers were constructed to find out how to design the CP-1 (Chicago Pile Number One), the world's first nuclear reactor. From the test piles they were able to determine the reproduction factor needed for the final reactor: the exact combination of ingredients and arrangements necessary for a nuclear chain reaction to be self-sustaining and efficient. (Argonne National Laboratory, Courtesy AIP Emilio Segrè Visual Archives)

THE PATH TO ATOM SMASHING

By 1938, when Otto Hahn and Fritz Strassmann first split the uranium atom asunder without being sure at all what they'd done, physicists and chemists had been peeling particles off atoms for several decades, always with surprising results. In 1919, after Ernest Rutherford had already discovered the nucleus by firing high-speed alpha particles at gold foil (see Chapter 1), he decided to try firing alpha particles through a tube of nitrogen gas. When he examined the results, he discovered something strange. In addition to his alpha-particle bullets (emitted naturally by radioactive radium), he found particles that had the same properties as hydrogen nuclei, even though the tube contained no hydrogen. (These particles later gained the name *proton.*) Rutherford concluded that he had struck a few nitrogen nuclei with the alpha particles and split the hydrogen nuclei away.

James Chadwick
(AIP Emilio Segrè Visual
Archives, W. F. Meggers
Collection)

The discovery of radioactivity had shown that certain atoms could disintegrate spontaneously in nature—but this was the first evidence that ordinary (nonradioactive) atoms were not indestructible. Over the next five years, with his colleague James Chadwick, Rutherford tried similar experiments skimming off protons from the nuclei of 10 different elements with alpha bullets. From these experiments they began to realize that the nucleus was not just a tiny positively charged entity; they were probing it, and if it could be probed, and chunks could be broken off in this way, it must have an internal structure.

Maybe, Rutherford hypothesized, there was a third atomic particle not yet discovered. Irène and Frédéric Joliot-Curie, in the tradition of Irène's parents, Marie and Pierre Curie, also were actively exploring atomic structure, in Paris. In their experiments they had been bombarding beryllium with alpha particles, and they were getting a strange particle they thought was a type of penetrating radiation. Chadwick tried a similar series of experiments, but he slowed the emitted particle with paraffin wax, which in turn shot hydrogen nuclei out of the paraffin. He was a tireless worker, spending day and night at the lab for three weeks straight. "Tired, Chadwick?" his colleagues would query. "Not too tired to work," came the rejoinder. When it was all finished, his first remark was, "Now I want to be chloroformed and put to bed for a fortnight."

During that time Chadwick succeeded in proving that he had discovered Rutherford's neutral particle. The neutron explained the discrepancy that no one had ever understood between atomic number and atomic weight. Carbon, for example, is 12 times as heavy as hydrogen, with an atomic weight of 12. But its atomic number is 6, based on the number of electrons it has. The negative charge of the electrons had to be evenly balanced by the protons in the nucleus. So where did all the weight come from, if it only had 6 protons in its nucleus? Now the answer was clear.

Meanwhile, in the United States, Ernest Lawrence had been working for several years on better ways to bombard atomic nuclei. Alpha particles had their problems. Since they carried a double positive charge, the positive charge of an atomic nucleus tended to repel them. In 1928 George Gamow had suggested that single protons—hydrogen ions—would work better, since they carry only a single positive charge. Several scientists worked on inventing a device for accelerating these particles to an effective speed in the following decade—including John Douglas Cockroft and Ernest Thomas Sinton Walton, who succeeded in devising the first particle accelerator in 1929, nicknamed the "atom smasher."

*Ernest Lawrence
holding the first
cyclotron in his hand*
(Courtesy of Lawrence
Berkeley Laboratory)

But the most successful was Ernest Lawrence. Lawrence's idea was innovative: Instead of jumping a particle up to speed with a single jolt, he conceived of sending it along a spiral path. As protons passed by each pole of a large magnet, their paths were deflected into circles, spiraling outward toward the rim of the device with ever greater speed. Finally, when the protons shot out of the instrument, they had accumulated a great deal of energy. Lawrence completed his first "cyclotron," as he called it, at the University of California in Berkeley in 1931. It soon became a key to future advances in the field of nuclear physics, especially as he built even larger and more powerful ones. Lawrence received the 1939 Nobel prize in physics for his work with particle accelerators.

and he even developed a system for shooting them first through paraffin to slow them down. Then when they arrived at their destination, they were moving so lazily that the nucleus snapped them up.

Now, in Chicago, his objective was to put together a test reaction that would move at a leisurely pace so that physicists could monitor it and prevent an explosion. He used naturally occurring uranium, ore that was mostly stable uranium-238. The structure he built alternated layers of uranium and graphite: uranium to promote the reaction, and graphite to slow the neutrons. He used 6 tons of uranium, 50 tons of uranium oxide and 400 tons of graphite blocks. Cadmium rods blocked the reaction from taking place until everything was in place.

On December 2, 1942, Fermi slipped the control rods out of the pile, and the chain reaction began. The unstable uranium-235 nuclei were split by the neutrons. Disintegration of the split atoms produced streams of more neutrons, which in turn shot out of the uranium block, into the graphite, which slowed them, and then into the next block, where they split more uranium-235 nuclei. Heat increased as the reaction continued. Although this was only a test and was never intended to produce power, Chicago Pile Number One was the world's first nuclear reactor. The atomic age had begun. And the path for the Manhattan Project was clear.

THE MANHATTAN PROJECT

Under great secrecy, the U.S. government assembled the top scientists in the country—many of the top scientists in the world, in fact—at a desert site called Los Alamos. Located in a remote area of north central New Mexico, Los Alamos was a wilderness perched atop a high mesa, and the only settlement was a private school for boys. The mesa rose 30 miles west of the rugged Sangre de Cristo mountains, the southernmost chain of the Rocky Mountains.

Those who arrived at Los Alamos were sworn to uphold strict security restrictions. The cream of American universities was brought together there: a community of mathematicians, scientists, engineers and chemists, whose job it was to determine how much material it would take and what arrangements would be most effective to produce the most lethal weapon the world had ever seen. Theirs, too, would be the job of designing and testing the devices that would bring the material together into an explosive unit. The idea was to figure out the quantities of ores and materials needed, to assemble them in the best way possible for the most effective chain reaction, and to devise a system of explosive devices to set it off.

The director of the project was J. Robert Oppenheimer, a tall, rangy young man known for his brilliance and charisma. At Los Alamos he built a community rare in its energy, intensity and focus. The job was top-secret and deadly serious. But for those who built the bomb it also was a giant

The Fat Man implosion bomb developed at Los Alamos   (Courtesy Los Alamos National Laboratory)

challenge, one that tested every fiber of their intellect, every power of reasoning and logic they possessed.

It took four years for the team at Los Alamos to design and build two types of bombs. One, called "Little Boy," was a uranium bomb that was triggered by a U-235 "bullet" that was impelled into a U-235 sphere by an explosive. The other was a plutonium implosion-type bomb called "Fat Man." It consisted of a plutonium core, surrounded by an initiator of polonium and beryllium and a circle of explosive wedges.

By July 1945, four bombs were completed: one plutonium assembly on a tower for a test; two others, one of each type, for possible use; and one plutonium type in reserve.

Ironically, Germany had surrendered to the Allied Forces in May 1945. What had begun as a focused push to stop the Germans and win a race against a German atomic bomb suddenly became unfocused for many of the scientists once the war with the Germans was over.

Throughout this period, Niels Bohr had argued for an international announcement about the project; at the very least he wanted the Russians to know about it. International control, he believed, was the only safe way to handle such a powerful bomb. But his overtures were treated with great suspicion by Roosevelt and British prime minister Winston Churchill.

In July 1945 World War II still was not over by any means. Japanese forces continued to fight bloody battles in what had become known as the Pacific Theater. The Japanese had an old saying, one long-time American correspondent in the area reported: "We will fight until we eat stones!" Americans began to believe that almost nothing would cause Japan to surrender.

Some of the scientists at Los Alamos had relatives fighting in the Pacific Theater. For them the situation called for ending the war as quickly as possible. But many of the scientists at Los Alamos, firmly believed that a warning should be given, an opportunity for surrender before the bomb was dropped.

HIROSHIMA AND NAGASAKI

On July 16, 1945, after several hours of delays because of bad weather and dangerous winds, the test of the first Fat Man at the Trinity test site, at Alamogordo in south central New Mexico, went exactly as planned. The first atomic bomb exploded over the desert. The scientists were ecstatic. The success represented, as one said, the best years of their lives.

What happened next, as Einstein once remarked, had a weird inevitability. On July 26, 1945, at 7:00 P.M., U.S. president Harry S. Truman released a document for dispatch known as the Potsdam Declaration; it was signed

Atomic test at the Trinity site   (Courtesy Los Alamos National Laboratory)

by Truman, the president of Nationalist China and the prime minister of Great Britain, and its transmission was picked up in Japan at 7:00 A.M. Tokyo time on July 27. The statement issued an ultimatum, requiring Japan's unconditional surrender and outlining the terms, concluding: "The alternative for Japan is prompt and utter destruction."

If Japan decided to continue the war, it stated, "the full application of our military power, backed by our resolve, will mean the inevitable and complete destruction of the Japanese armed forces and just as inevitably the utter devastation of the Japanese homeland."

The next day, Japan's prime minister, Baron Kantaro Suzuki, held a press conference in response to the Potsdam document, during which he announced that his government found no important value in it, stating, ". . . there is no other recourse but to ignore it entirely and resolutely fight for the successful conclusion of the war."

For decades afterward debates would continue about the details of this exchange, but the United States had sent its warning, and the clear message received back was that Japan intended to keep fighting. At that time, Fat Man and Little Boy had already traveled from New Mexico to staging sites in the Pacific.

On August 6, 1945, an American aircraft called the *Enola Gay* dropped a Little Boy atomic bomb on Hiroshima, Japan. As the crew of the *Enola Gay* looked back on the city, it seemed to disappear entirely, engulfed in billows of smoke and fire. One crew member later said, "I don't believe anyone ever expected to look at a sight quite like that. Where we had seen a clear city two minutes before, we could no longer see the city." Robert Caron, the tail gunner, had the clearest view as the plane pulled away:

> *The mushroom itself was a spectacular sight, a bubbling mass of purple-gray smoke and you could see it had a red core in it and everything burning inside. As we got farther away, we could see the base of the mushroom and below we could see what looked like a few-hundred-foot layer of debris and smoke . . .*

Four square miles of the city were wiped out entirely, leveled to the ground, and 90 percent of the city's buildings were destroyed by the equivalent of 12,500 pounds of TNT. Within a mile of the explosion's center the temperature rose as high as 1,000° F, leaving charred human flesh and melted metal in its wake. The death toll could not immediately be calculated.

No word of surrender came from the Japanese. On August 9 another bomb—Fat Man—was dropped on the southern city of Nagasaki, killing 40,000 and destroying the city. Finally, on August 14, 1945, over the objections of his military commanders, Emperor Hirohito announced

J. Robert Oppenheimer directed the scientific team at Los Alamos
(Los Alamos National Laboratory, Courtesy AIP Emilio Segrè Visual Archives)

Japan's surrender. On September 2, World War II officially came to an end as surrender terms were signed aboard the battleship USS *Missouri*.

By the end of 1945, the deaths in Hiroshima would mount to 145,000, and many, many more were injured. In the next five years tens of thousands more would die of radiation poisoning, a lingering effect of the devastating bombs.

AFTERMATH

When the news first came in at Los Alamos that the atomic bomb had worked, most of the scientists felt a high sense of jubilation. Four years of work had met with success. The many intricate devices they had designed and tested had done their job. A long and costly war would now end as a result of their efforts. But unease also filled the spare wooden structures at Los Alamos, and dismay mounted as more detailed reports came in. For many of the scientists who rejoiced in August 1945, a lifetime of questioning would follow, as they were haunted by the death and destruction caused by their creation. Oppenheimer remained tortured for the rest of his life by the horrible potential of the bomb he had helped create. In his last public speech at Los Alamos he warned:

> *If atomic bombs are to be added to the arsenals of a warring world, or to the arsenals of nations preparing for war, then the time will come when mankind will curse the name of Los Alamos and Hiroshima.*

> *The peoples of this world must unite, or they will perish. This war, that has ravaged so much of the earth, has written these words. The atomic bomb has spelled them out for all men to understand. Other men have spoken them, in other times, in other wars, or other weapons. They have not prevailed. There are some, misled by a false sense of human history, who hold that they will not prevail today. It is not for us to believe that. By our works we are committed, committed to a world united, before this common peril, in law and in humanity.*

It was the beginning of an impassioned and fruitful search for peaceful uses for atomic power. A group known as Atoms for Peace was formed by a band of physicists, headed by Niels Bohr, who believed that atomic power should never again be used in this way and who campaigned fervently to see that humanity should learn important lessons from the events of August 1945. At the heart of the most sophisticated planning for avoidance of nuclear holocaust, many of the world's finest scientists can be found today, as they continue to promote the humane use of their discoveries and technology.

PART TWO

THE LIFE SCIENCES

THE GROWTH OF MICROBIOLOGY AND BIOCHEMISTRY

*I*f figuring out how the world works in terms of physics, chemistry and astronomy seemed complex, the realm of living things held even more uncanny mysteries. While chemists and physicists delved deep into the atomic and subatomic realms, life scientists continued on a similar path—a path that sought answers about the nature of living things and how they functioned, and, ultimately, about the basis of life. The intricacies of even the simplest living creatures (many of which are too small to see with the naked eye) had baffled everyone for centuries—from the Greeks to William Harvey and Antony van Leeuwenhoek in the 17th century to Louis Pasteur and Robert Koch in the 19th. And the age-old questions all remained far from answered at the beginning of the 20th century: What are living things? What makes them different from rocks or dirt or stars? What sustains them? How do they function and what are the processes that are inherent to life in different organisms?

For centuries, investigators had been trying to find out more about how organisms function, mostly by working from the outside in. Among the ancients, thinkers like Aristotle and Pliny focused at first on the morphology, or outward shape, of organisms. Later, scientists such as William Harvey who first applied the principles of observation and experiment to living things began to look at the way organs and systems of organs—such as the circulatory system—worked within these organisms. Then came the realization that organs are made up of tissues, and in the 19th century Matthias Jakob Schleiden and Theodor Schwann recognized the function of a boxlike structure they called the "cell," of which all tissues, and thus all organs and all creatures, plant and animal, are made. With the approach of the 20th century, experimental technique and technological breakthroughs began to take on ever greater importance for biologists, as they, like their colleagues the physicists, moved more and more toward the very small.

A QUESTION OF NERVES

For a long time, scientists had been trying to figure out what the structures called "nerves" were and how they worked. Initially, they seemed like strange, mysterious threads, and up until the 18th century, scientists assumed they were hollow tubes, like veins and arteries, and carried a fluid or "spirit." The Swiss physiologist Albrecht von Haller (1708–77) finally introduced a more experimental approach, which led to a recognition that nerves play a key role in the interplay of stimulus and response. And Haller also noticed that all nerves lead to and from the brain and spinal cord, which, he surmised, act as centers of sense perception and responsive action.

By the end of the 19th century, the introduction of the cell theory had provided new insights to apply to nerves and how they work. Before that, biologists had discovered structures in the brain and spinal cord, but they weren't at all certain what these were, what their purpose was or how they

Santiago Ramón y Cajal, the founder of neuroanatomy (Parke-Davis, Division of Warner-Lambert Company)

Minute organisms, such as Pasteur and Koch's bacteria, they found, were easier to study and helped provide insights about the basis of life—a key question around which all biological inquiry kept circling.

---

worked. Then in 1891 German anatomist Wilhelm von Waldeyer (1836–1921) announced his view that nerve fibers were part of the nerve cell—that they were fine, extended fingers leading to and from the nerve's center. Called the "neuron theory," von Waldeyer's work also indicated that these extensions often closely approached each other but didn't touch, creating a gap that later came to be called a synapse. And the first to substantiate this theory, coordinating the idea of the cell as the basic unit of living matter with the idea of systems in organisms, was Santiago Ramón y Cajal of Spain.

When Ramón y Cajal (1852–1934) graduated as a doctor of medicine from the University of Madrid in 1877, no one had yet come up with a stain that could make the parts of the nervous system stand out against the surrounding tissue (known as neuroglia). Then, in the 1880s, he heard about a stain recently developed by Camillo Golgi (1844–1926), an Italian pioneer in the study of the minute structure of the nervous system. Robert Koch, Paul Ehrlich and others had recently brought the staining of cells into prominence—but these men were using synthetic dyes. In 1873 Golgi [GOHL jee] had announced a method of using silver salts that produced spectacular results, especially for observing nerve cells. Using the new stain, Golgi had already confirmed the tiny gap known as the synapse.

But Ramón y Cajal [rah MOHN ee kah HAL] improved on Golgi's stain and, based on his meticulous studies, established the neuron theory beyond question. Often referred to as "Don Quixote of the Microscope," he completely disproved the idea, widespread at the time, that gray matter of the brain and spinal cord was composed of a continuous, interconnecting network, one branch flowing into the next like tributaries to a river or veins on a leaf. He succeeded in tracing long nerve fibers that never fused with any other. With his microscope and scrupulous experimental technique he confirmed that the fundamental unit of the nervous system is the neuron, with its branching processes and its axon. And he also put forth the hypothesis that the nerve impulse passes in only one direction as it passes from neuron to neuron.

Ramón y Cajal was awarded the Nobel prize in 1906 (which he shared with Golgi) in physiology or medicine. His work established the new field of neuroanatomy, one of the many new disciplines being formed as a result of advances in microscopy and staining in the late 19th and early 20th century.

A great, ongoing controversy had raged since the time of Newton, when life scientists had first tried to apply the same mechanistic concepts that worked so well in physics to the world of living things. Many life scientists objected to this purely materialistic approach. Life, they felt, was different from brews of chemicals and assemblages of levers and valves. Something more seemed to be present, a sort of "vital" force. And so, their point of view came to be called "vitalism." In 1895, the life science world was split. Are living things special in some way? they wondered. Do they possess a life-sustaining "essence," a soul, or life force? Or are they merely assemblages of atoms and molecules, like inanimate substances, subject to all the same physical laws as are a desk, a wagon or a meteorite? To many, the latter seemed an impossible sacrilege.

BUCHNER'S BREW

In his textbook of organic chemistry, the first volume of which was published in 1861, Friedrich August Kekulé von Stradonitz had defined organic compounds simply as those containing carbon. He made no mention of a life force or any other distinguishing characteristic. For the first time, organic substances began to be examined not in terms of the presence of a life force, but in terms of chemical elements, described in much the same way as any other substance.

Many people found Kekulé's approach unsettling, and factions lined up on either side of the issue. Then in 1897, just as the 20th century was about to dawn, German chemist Edouard Buchner (1860–1917) tried an experiment. Fermentation had long been regarded as one of the life processes, a chemical reaction that could only occur in the presence of living cells. So Buchner [BOOKHner] collected a group of yeast cells, known to be associated with fermentation, and ground them up with sand so that absolutely no living cells could remain. Then, just to be doubly sure, he passed the ground-up substance through a filter to obtain a cell-free juice.

What happened next was not at all what Buchner expected. He was sure that, in the absence of cells, no fermentation would take place. He carefully kept the juice he'd concocted free of contamination by any living cells—otherwise his experiment would not be a good test. Then he added a concentrated sugar solution, which was recognized as a good way to preserve against contamination by microorganisms. To his amazement, the cell-free yeast juice and sugar mixture began to ferment! What everyone thought was a life process had taken place in an absolutely nonliving mixture. Buchner went further. He killed yeast cells with alcohol and found that dead cells ferment sugar as readily as living ones do. These results were both startling and exciting, and in 1907 Buchner received the Nobel prize in chemistry for his work.

Vitalists (even Buchner himself) had been sure that all this was impossible. Yet the conclusions were clear: "Ferments," as agents of fermentation were called at the time, were in reality dead substances that could be isolated from the living cells in which they were usually found. And these substances could be made to do their work in a test tube in a laboratory.

It's now generally accepted that life follows the same laws that govern the nonliving world. But in 1897, Buchner's little experiment was in a way revolutionary for those studying living organisms. From his work, biologists and chemists gained confidence that problems in biology were not innately beyond the scope of laboratory examination and understanding. Like the phenomena of the inanimate world, life processes would yield up answers in response to scientific experiment and observation, even in the absence of life. The stage was set for the mechanistic study of the chemistry of cells.

Philosophical battles over vitalism continued, certainly. In 1899, German naturalist Ernst Heinrich Haeckel (1834–1919) published the view that the mind, though the result of creation, depends on the body and does not live beyond its death. (He was also the first to use the word *ecology* to describe the study of living things in relationship to each other and to their environment.) For many, this contention strayed too far from the evidence at hand and countermanded traditional views too strongly. For others, it seemed to make sense. The evidence was piling up against the vitalists' viewpoint, but throughout the century, scientists would continue to seek answers to the questions, What is life? and, How did life begin?

BODY CHEMISTRY

With his breakthrough experiment of 1897, Buchner had set the stage for much fruitful experimentation in the life sciences, essentially establishing a new field that combined both chemistry and biology: the field of biochemistry. Now the chemistry that took place inside cells could be studied in the laboratory, in test tubes, outside cells, because Buchner had shown that the cell itself contributed nothing special to the reactions that took place there. With this experiment, the field of endocrinology—the study of ductless glands and their secretions—was founded.

In 1901, Jokichi Takamine, of Japan, discovered a substance called adrenaline (epinephrine), which is secreted by the adrenal gland, an endocrine gland located near the kidneys. Takamine not only isolated this substance that constricts blood vessels and raises blood pressure, but, in another blow against the vitalist viewpoint, he also succeeded in synthesizing it.

William Bayliss (1860–1924) and Ernest Starling (1866–1927), two phys-
iologists in England, were meanwhile experimenting with the pancreas, a
large, soft digestive gland. They found that, even though they severed all
the nerves leading to the pancreas, it still secreted digestive juice as soon as
the stomach's acid and food contents emptied into the small intestine. In
1902 they succeeded in isolating the cause. Acid in the small intestine, they
found, caused secretion of a substance from its walls, which they called
secretin. This substance, carried through the blood to the pancreas, triggered the digestive juices. Secretin, Bayliss and Starling realized, and the
epinephrine discovered by Takamine, were part of a system of chemical
messengers, which they called hormones (from the Greek word *horman*,
meaning "to urge on"). These chemical messengers were special proteins
secreted by glands in one part of the body and carried by the blood to
regulate reactions in specific cells in other parts of the body. Bayliss and
Starling's work established the importance of hormones, and their new
hormone theory proved to be enormously fruitful, paving the way to a
method for treating a deadly disorder that had plagued—and killed—humans since antiquity.

Diabetes, meaning "passing through," was the name the ancient Greeks
gave this disorder because the enormous amounts of water the sufferer drank
seemed to pass right through the body. The Romans added the word
mellitus, meaning "honey," because the urine was abnormally sweet—so
sweet that it drew flies. Once the disorder set in, death always followed.
There was no cure and no way to slow the process.

By 1920, it was also known that the sweetness of diabetic urine was caused
by abnormally high glucose levels, which showed up in the blood of diabetic
patients as well. Also, when the pancreas was removed from animals under
experimentation, the animals developed a condition very much like diabetes.
So, based on the work done by Bayliss and Starling and these new findings,
suspicion was high that the disorder was caused by the lack of a hormone
secreted by the pancreas and needed to regulate the glucose levels in the
blood. Without it, glucose built up and diabetes set in. The hormone, which
no one had ever found, even had a name: *insulin*.

Though some people might have thought him an unlikely candidate to
solve this age-old problem, Frederick Grant Banting (1891–1941) thought
he had an idea. Recently graduated from medical school and just returned
from military duty, Banting was a young Canadian physician just starting
out in practice and teaching part-time at the University of Western Ontario
Medical School. While preparing his lecture notes one day in 1920, Banting
came across an item in one of the scientific journals that intrigued him. If
the pancreas was ligated, or tied off, from the intestine so that it could not
send its digestive juices through its ducts, the article said, it would shrivel
up. Excitedly, he jotted a note to himself: "Ligate the pancreatic ducts of

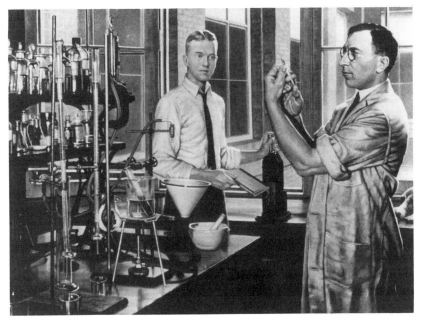

Charles H. Best (left) and Frederick Banting, discoverers of insulin (Parke-Davis, Division of Warner-Lambert Company)

dogs. Wait six to eight weeks for degeneration. Remove the residue and extract." In this way, Banting reasoned, he could isolate insulin from the shriveled pancreas, free of destructive digestive juices. But Banting had no resources for research, and no laboratory.

He headed for the University of Toronto, where John Macleod, an expert on diabetes, was head of the physiology department. Banting outlined his idea and asked for laboratory space for an eight-week period. But Macleod turned him down. And Macleod turned him down again the second time he asked. However, the third time worked. Macleod finally agreed to let Banting use his lab while he was on vacation and even suggested the name of a young student just entering medical school who might be interested enough to be Banting's assistant. Charles H. Best (1899–1971) agreed eagerly to the plan. Though he usually earned money playing professional baseball in the summer, he thought he could get by this once on discharge pay he'd just received from the army.

Best and Banting used dogs for their experiments, but they always treated them kindly, with affection and care. Tying off the pancreas required an operation, but Banting used an anaesthetic and watched over their recovery with the same concern he would show a human patient. Unfortunately, the first operation didn't take: The catgut they used to tie off the pancreatic ducts disintegrated, and the dog never developed any symptoms, but several weeks passed before they realized the mistake. Money, meanwhile, was running out. So Banting sold his old, dilapidated Ford to buy food, both for the experimenters and their dogs. But finally, Marjorie, a dog whose pancreas they had tied off with silk thread, developed symptoms of diabetes in late July 1921. The two young scientists removed her pancreas, which had shriveled, just as expected, and they ground it up and dissolved it in a salt solution. They administered the brew to Marjorie, and all the symptoms of diabetes disappeared.

When Macleod came back from his vacation, he was amazed. Banting and Best had isolated insulin. Macleod and a fourth man, James Collip, were added to the team to purify and standardize the hormone, and Banting and Best presented their discovery at a scientific meeting in November 1921. But, in the flurry that followed, Macleod, as head of the department, got the credit in many reports, and in 1923, when the Nobel prize was awarded for

Best and Banting with Marjorie, the dog who helped them prove the existence of insulin (Eli Lilly and Company)

An early insulin research laboratory (Eli Lilly and Company)

the achievement, it was given not to Banting and Best, but to Macleod and Banting. Banting was furious that Best was not included, and Macleod thought that Collip's work should have been recognized. So, as has often happened with the Nobel prize, the recipients redistributed the $40,000 they each received to share it with their colleagues. Manufacture of insulin went swiftly into production to meet the demand from doctors for this new, and for the first time, effective, treatment for diabetes.

A SMALL WORLD

Traditionally, questions in the life sciences have arisen out of the most practical of concerns: keeping ourselves alive and well. So the study of medicine—usually thought of as an "applied science"—travels hand-in-hand with theoretical advances, often leading the way. And in the first half of the 20th century, as much as or more than ever, the search for health drove the search for knowledge.

In the 1790s, Edward Jenner had introduced the first vaccine when he came up with the idea of innoculating a healthy boy with fluid from a blister of a girl sick with cowpox, a disease similar to but milder than smallpox. Then, in one of the riskiest experiments in the history of medical research, he exposed the boy to smallpox—and the boy didn't get sick. (The boy could easily have died, though, if Jenner's premise had been incorrect. Smallpox was a deadly disease that killed one out of three people in Europe during its worst epidemics.) Jenner's experiment worked, resulting in development of the first effective vaccine, and Jenner became a hero.

But no one knew in those days exactly why the smallpox vaccine worked or what really caused that, or any other, disease. For the most part, physicians could do nothing but treat symptoms, let an illness run its course and console the survivors of those who died.

Finally, Louis Pasteur made a major breakthrough in the 1860s with the introduction of his "germ theory," which identified tiny organisms as the agents that cause communicable disease. And in 1876 Robert Koch found the germ—a bacterium, *Bacillus anthracis*—responsible for the dread anthrax disease that wiped out whole herds of domestic animals and often also affected humans. It was the first definitive connection made between organisms so small they could be seen only with a microscope (microorganisms, as they are often called) and illness.

By the 1890s, several bacteria had been identified and linked with diseases and infections, and new methods of killing them or at least controlling their spread had been introduced into hospitals and surgical procedures. But some diseases continued to evade explanation and seemed harder to fight. Rabies was one, and Pasteur had conjectured that some minute organism too small even to be seen through a microscope was responsible. Another disease that resisted explanation was a malady that plagued tobacco plants, and as early as 1892 it had been suggested that the disease was caused by something that could be passed through the finest of filters.

That's where a Dutch botanist named Martinus Beijerinck (BY er ink) came in. Beijerinck (1851–1931), the son of a tobacco dealer, had training both in botany and chemistry. In 1895 he tried an experiment. Pressing out the juice of tobacco leaves infected with mosaic disease, he examined the residue carefully for suspicious bacteria. He could find nothing. He tried growing a culture of culprit bacteria from the juice, but nothing grew. Yet a healthy plant, when exposed to the juice, would become infected with mosaic. What was causing the infection, if no bacteria were present? He passed the juice through a filter so fine that it would remove any known bacterium. Still it infected healthy plants.

Maybe the disease was caused by some sort of toxin. But no, Beijerinck found that he could pass the disease from plant to plant to plant—so whatever caused it, he concluded, must be growing and multiplying.

In 1898, after considerable testing, Beijerinck published his observations, announcing that tobacco mosaic disease was caused by an infective agent that was not bacterial. He called it a filterable virus (*virus*, from the Latin word for "poison"). And so he discovered a class of infectious agents that would later prove to be the cause of numerous ailments of plants and animals, including such illnesses among humans as yellow fever, polio, mumps, chickenpox, smallpox, influenza and the common cold. Not until a generation later, however, did biologists even begin to understand the structure of the virus.

Finally in 1935 Wendell Meredith Stanley (1904–1971), an American biochemist, made a breakthrough. Operating on the hunch that the tobacco mosaic virus was really a protein molecule, he mashed up a lot of tobacco leaves that he had infected with the disease and used basic crystallization methods that had worked on other proteins. At last he succeeded in obtaining a group of fine crystals shaped like needles. He isolated them and found that their ability to infect exactly matched the infective properties of the virus.

This proved to be a tough piece of information for many people to swallow: viruses were alive, weren't they? They could reproduce themselves within cells—one of the key criteria distinguishing living things. But Stanley had crystallized what apparently was a virus in the same way scientists crystallized inanimate chemicals. This news seemed to place viruses in a never-never land between the living and nonliving. It was a confusing and unsettling idea, and controversy raged as people kept trying to classify the virus as one or the other. And once again the old argument was raised about what life is and what it isn't.

The tale began to spin out further as research in the 1940s showed that viruses contained both protein and nucleic acid. Solutions of nucleic acid alone, it soon became apparent, could change certain physical characteristics of bacterial strains. And for the first time biochemists began to look at nucleic acids as the possible carriers of genetic information, as will begin to emerge in Chapter 7, "Pursuing the Trail of Genetics and Heredity." Now research on viruses and the study of heredity had begun to merge—a trend that would bear more fruit after World War II, in the latter half of the century.

In the meantime, in the first half of the century, researchers focused a great deal of effort on understanding how tiny microorganisms functioned and how to defeat their life functions, and so overcome the diseases they caused. In the process they found out a lot about life functions in general.

PAUL EHRLICH AND THE "MAGIC BULLET"

"As long as I have a water tap, a flame and some blotting paper," Paul Ehrlich [AIR lik] once claimed, "I can work in a barn." For really quality thinking, though, he also seemed to require enormous quantities of mineral water and endless boxes of cigars, which he smoked incessantly. In fact, cigars were so necessary to his thought processes that he would tuck an extra box under his arm whenever he went out—just to be sure he had some on hand.

Paul Ehrlich (1854–1915) was sometimes hard to get along with and was always sure he was right. He used to try the patience of his assistants by giving them stacks of index cards each morning with detailed instructions for the day's experiments. And variation from the instructions was not greeted warmly.

But Ehrlich got things done, and, because of that trait, coupled with brilliance and intuition as an experimenter, he made an enormous contribution both to science and human health as the founder of chemotherapy.

As a young scientist still in medical school at the University of Leipzig in Germany, Ehrlich became intrigued (as Camillo Golgi and Robert Koch had before him) with the way the new aniline dyes could make various microscopic structures easier to examine. While still in school he discovered several useful bacterial stains, and he wrote his thesis on the subject. Not everyone was convinced he would go far with it, however. Koch was visiting Ehrlich's school one day and met the young enthusiast. "Very good at staining," Koch later pronounced, "but he will *never* pass his examinations." But Ehrlich passed.

And, after obtaining his medical degree in 1878, Ehrlich discovered a good method for staining the tubercle bacillus, Koch's area of interest. This feat brought him to Koch's attention once again, and the two men worked together between 1882 and 1886, when, unfortunately, Ehrlich caught a mild case of tuberculosis and left for Egypt to recover.

In 1889, when Ehrlich returned, he went to work with Emil von Behring (1854–1917) and Japanese bacteriologist Shibasaburo Kitasato (1856–1931), both of whom had also worked with Koch, and Ehrlich obtained a position at the University of Berlin in 1890. Scientists were gaining insight into the causes of disease and how natural substances in the blood worked to produce natural immunities. Biologists began to think maybe these natural immunity builders were part of the key to what made vaccines work: Vaccines provided an ally for the body's natural immune system. In that same year, 1889, Kitasato and Behring announced their discovery that when they injected animals with increasing doses of tetanus toxin, too small to be lethal, the animals' blood developed substances (antitoxins) that neutralized the toxin, or poison, produced by the tetanus bacillus. They also found that they could take the liquid part of the blood (known as the serum) of an animal immunized in this way and use it to immunize other animals. This simple step could be used to prevent illness from resulting from what would otherwise be a fatal dose of a toxin or bacteria. (Kitasato, by the way, had already isolated the bacillus causing tetanus and another causing anthrax while working with Koch in 1889. And, after returning to Japan, he isolated the agent that caused bubonic plague during an outbreak in Hong Kong in 1894.)

Meanwhile, Behring, Kitasato and Ehrlich were looking for a cure for diphtheria, a deadly disease that attacked children in particular and was usually fatal. They'd noticed that children who caught diphtheria and survived seemed to be safe from catching it again in adulthood. Apparently, in the fight against the disease, the child's body had built up antibodies that remained in the blood and protected against future attacks. But the risks of gaining immunity in this way were too high. The three bacteriologists used the same methods as with tetanus, with Ehrlich working on the technical aspects of dosages and techniques of treatment. They tried out their new diphtheria antitoxin in the diphtheria epidemic of 1892, and it worked. For this work, Behring, who had come up with the idea, received the Nobel prize in physiology or medicine in 1901.

After their success with diphtheria, Ehrlich quarreled with Behring, and Kitasato returned to Japan. So Ehrlich struck out on his own. Impressed with the work on the diphtheria antitoxin, the German government set up an institute for research on serum and made Ehrlich its head. Not only did Ehrlich excel in the experimental side of research, conceiving of ingenious methods and tools for testing hypotheses and inventing useful laboratory techniques, but he was never satisfied with just finding a cure. He wanted to know *how* the diphtheria toxin attacked and how the antitoxin worked to keep the toxin from injuring the cells of the body. He wanted to know the chemistry behind the events he saw. Returning to his early interest in stains, he thought about the fact that a stain's value lay in its ability to make a cell structure stand out, or to color a bacterium so that it could be seen against a colorless background. There had to be a chemical explanation for this phenomenon. The stain must combine with some substance in the bacterium, usually with the result that it killed the bacterium. Maybe this fact could be used to fight bacteria. In fact, maybe a dye could be found that stained—and killed—harmful bacteria without harming the normal cells of the body. Maybe a "magic bullet" could be created that would in effect target the bacteria attacking the host—seeking out the parasites and destroying them. And so the field of chemotherapy was born.

Ehrlich set out to look for dyes that would stain and kill specific targets, and he found one, called trypan red, that he discovered could be used to destroy trypanosomes, a genus of protozoan (one-celled animal) that causes several illnesses, including sleeping sickness.

He began to suspect that the nitrogen atoms in trypan red were the key to its interference with the parasite's metabolism, and he thought of trying various compound combinations with arsenic to see what other magic bullets he might be able to come up with. Arsenic has many properties in common with nitrogen but is considerably more toxic, so this line of approach looked promising. He set everyone in his laboratory to work trying all the organic compounds containing arsenic—either natural or synthesized—that they could conceive of. They tried hundreds. By 1907 they had reached number 606; they tried it on trypanosomes without much success, set it aside with all the others, and went on.

Ehrlich received the 1908 Nobel prize in physiology or medicine, which he shared with Russian bacteriologist Ilya Ilich Mechnikov (who by that time was living in Paris), for work in the study of immunity. But, in fact, Ehrlich's greatest contribution was yet to come.

The following year, one of his associates, Sukehachiro Hata, was reviewing techniques for testing the effectiveness of arsenic compounds, and he happened to use number 606. To everyone's amazement, although 606 had no special effect on trypanosomes, he found that it did do considerable damage to spirochetes, which cause syphilis. Excited by his associate's report, Ehrlich confirmed it, renamed the drug "salvarsan" and announced the find in 1910. The magic bullet had been found to control one of the world's most destructive diseases, an illness that was often kept secret because it was spread through sexual intercourse. In its victims it caused sterility and, ultimately, led to paralysis, insanity and death. Ehrlich distributed some 65,000 units to doctors all over the world, without charge,

Paul Ehrlich and his associate Sukehachiro Hata working in the laboratory (Parke-Davis, Division of Warner-Lambert Company)

believing that eradication of this disease was more important than any income he should derive from its use. The discovery of salvarsan, now known as arsphenamine, marked the beginning of modern chemotherapy, the production of drugs that acted, in effect, as synthetic antibodies that would seek out and destroy invading microorganisms without harming the host.

Ehrlich often said he was a firm believer in the importance of the "four big G's" for success: Geduld, Geschick, Geld, Glück (patience, ability, money and luck). But when congratulated for his discovery of drug number 606, he said simply, "For seven years of misfortune I had one moment of good luck."

The comment underplays the tremendous amount of work involved—for Ehrlich was not only a driving taskmaster, but a prodigious worker himself. During the years between 1877 and 1914, Ehrlich published 232 papers and books. In addition, the experimentation itself was enormously laborious, as his secretary, Marthe Marquart, once explained:

> *No outsider can realize the amount of work involved in those long hours of experiments that had to be repeated and repeated for months on end. People often refer to 606 as the 606th experiment; this is not correct, for 606 is the number of the substance with which, with all the previous ones, very numerous experiments were made. The amount of work which all of this involved is beyond imagination.*

The techniques in chemotherapy developed by Paul Ehrlich have continued to bear fruit up to the present, and the cures he and his associates found for sleeping sickness and syphilis provided the best defense against these illnesses until the 1930s, when two more breakthroughs added to the arsenal scientists were rapidly mustering against disease.

SULFA, THE "WONDER DRUG," AND PENICILLIN

By the mid-1930s, laboratories all over the world were looking for dyes or other chemicals that could attack bacterial infections even more effectively, and private drug companies set up their own experimental laboratories to see if they could win the race. In Germany, Gerhard Domagk (1895–1954) became director of such a laboratory for the I. G. Farben Company, where he and his colleagues began working on the powerful streptococcus bacteria that cause blood poisoning. Domagk [DOH mahkh] began a systematic battery of tests on several newly synthesized dyes, and in 1932 he came upon one, an orange-red dye called Prontosil, that cured streptococcus infections experimentally in mice. This was exciting news, for this type of bacteria was smaller and more resistant to attack than Ehrlich's syphilis spirochetes had been.

Even before Domagk had a chance to test his find on humans, a physician appealed to him to help a baby that was dying of staphylococcal blood poisoning. Prontosil at that time had been tested only on streptococcus infections, not on staph infections. But the doctor convinced Domagk to let him try it; nothing else was going to save the little boy. Four days after the baby received Prontosil, his temperature dropped, and within three weeks he had recovered entirely. Domagk's own young daughter, Hildegard, was also cured of a streptococcus infection in February 1935, and the drug gained worldwide fame when it saved the life of Franklin D. Roosevelt, Jr., the son of the U.S. president, who also had developed a dangerous infection.

The active ingredient in Domagk's Prontosil, it turned out, was sulfanil-amide, and a constellation of related organic compounds was soon found—known as sulfa drugs—that proved highly effective against streptococcus, gonococcus (cause of the venereal disease gonorrhea), meningococci (cause of meningitis), and some types of pneumococci, staphylococcus, brucella and clostridium infections.

René Jules Dubos (1901–82), an American bacteriologist who had a long and productive career, showed early in his lifework that natural substances produced by microorganisms could also serve as antibacterials. In 1939, Dubos [du BOZ] derived substances from soil bacteria that proved effective against the pneumococcus. And this discovery quickly led to a reexamination of an event that had occurred back in 1928, a discovery made in the Inoculation Department at St. Mary's Hospital in London by a bacteriologist named Alexander Fleming.

If Fleming (1881–1955) had been a more organized experimentalist and a less observant scientist, the world might never have enjoyed the benefits of one of the most powerful agents against disease ever discovered. One day in 1928, after returning from a vacation, he was cleaning up a batch of bacterial culture dishes that he'd left stacked in a corner of his lab. But as he was piling them all in a bath of disinfectant to destroy the cultures and prepare the dishes for reuse, he happened to notice something strange about one of them, and he snatched it back from the pile in the sink.

What attracted his attention was an unusual mold growing in one area of the culture dish, surrounded by yellow colonies of staphylococcus. Spores of all kinds filled the hot summer air of the old lab he worked in, and so the presence of a mold was not in itself that striking. What was striking was that all the bacteria in a one-inch arc surrounding the mold were completely colorless and translucent. Obviously, to the trained eye of a bacteriologist, something had killed the staphylococcus surrounding the mold. Fleming knew he had found something. He took a photograph, scraped off some of the mold to allow it to reproduce, and preserved the dish. He grew samples of the mold and sent it to other laboratories. And he and his colleagues studied it.

The mold was *Penicillium notatum,* and it produced a substance that Fleming found acted as a good antiseptic for test tubes and culture plates and was useful to purify strains of bacteria. He and his laboratory assistants found that penicillin was effective against the germs that cause scarlet fever, pneumonia, gonorrhea, meningitis and diphtheria, but they didn't succeed in purifying enough to test its usefulness as a medicine. And so, except for laboratory uses, Fleming's mold sat on the shelf for 10 years.

Then in 1938, two bacteriologists from opposite ends of the world—Ernst Chain (1906–79), a Jewish émigré who had fled Nazi Germany, and Howard Florey (1898–1968), from Australia—got together at Oxford University in England and came across penicillin in a survey they were doing of all the scientific literature on antibacterial agents. They found that a strain of the mold (offspring of one of the samples sent out by Fleming) existed right in their own lab, and they were off and running. Florey and Chain and their coworkers at Oxford figured out how to mass-produce penicillin so that they had enough to test on human patients. As World War II began, ever greater urgency existed for completing the work and making the new antibacterial available, and the work was transferred to pharmaceutical laboratories in the United States.

Fleming's particular strain of the penicillin mold was never found again growing outside a laboratory, although a similar mold (the strain that is now used) was finally found in Illinois in 1943, after an intensive search. So if it hadn't been for Fleming's watchful eye and quick hand, penicillin might have just washed down a drain that day in his lab. Instead, thousands of soldiers were saved from death from infections during the war, and since then, such diseases as pneumonia and scarlet fever have almost completely lost their ability to seriously threaten lives. In December 1945, Fleming, Florey and Chain were awarded the Nobel prize in physiology or medicine for their work.

With the discovery of penicillin, laboratories all over the world were sparked to search soil funguses for still more antibiotics, and many were found. A cure for Rocky Mountain spotted fever and for typhus was found. And Selman Waksman, who worked for a U.S. pharmaceutical manufacturer and taught at Rutgers University, discovered streptomycin in 1943, produced by a mold that one of his students found in a chicken. It was the first antibiotic (a word coined by Waksman) found that absolutely wiped out the tubercle bacillus. Many had weakened the bacillus, but Waksman's streptomycin was the first to kill it. In an unusual move, Waksman's employer, Merck and Company, decided to make the new drug generally available without patenting it, considering it to have such widespread humanitarian importance that it should be manufactured and distributed as widely as possible.

The results of all these breakthroughs were dramatic. Deaths from pneumonia and influenza in the United States dropped off 47 percent between 1945 and 1955, with mortality from syphilis dropping off 78 percent. Not all children had been vaccinated against diphtheria by this time, but mortality from that disease decreased by 92 percent. Drugs from the penicillin family and the scores of other antibiotics available quickly made death from infectious disease extremely rare wherever these resources were available—whereas before the beginning of the 20th century death from infectious disease had been the major cause of all deaths.

A MATTER OF DIET

During the last half of the 19th century, Pasteur's germ theory and, later, the discovery of viruses as agents of disease, had caught the imagination of most of those engaged in solving problems of public health. But another key factor also began to emerge from studies and observations. Since the 18th century, bouts of scurvy among shipboard explorers had been wiped out almost entirely by introducing lime juice into the diet of British sailors. Disease, it seemed, could also be caused when certain required substances were missing from an organism's diet. In the field of medicine, the interest focused on the diet of the human organism.

During the 19th century, protein was found to play an important role in diet, and a distinction was made between "complete" proteins, which could support life when present in the diet, and "incomplete" proteins, which could not; but no one knew exactly what the difference was. By 1820, scientists had isolated a substance called glycine, a simple molecule present in the complex gelatin molecule (a protein). Glycine belonged to a class of compounds called amino acids, which are an essential component of proteins. Soon other amino acid molecules were also found in proteins, and by 1900 a dozen different amino-acid building blocks had been discovered.

An English biochemist named Frederick Gowland Hopkins (1861–1947) was the first to show that not all proteins contain all amino acids, and that some amino acids are essential to life while others are not. In 1900 he found a protein isolated from corn that contained no trace of the amino acid tryptophan, and this protein, called zein, could not support life. Then he added tryptophan to his zein brew, and amazingly now zein could support life. In the early 20th century, other experiments showed that, while the body could manufacture some amino acids, others, known as "essential amino acids," had to be supplied through nutrition. Without them, sickness and, eventually, death, would ensue.

So diet—not just vanquishing germs—was important to health. But amino acids definitely were not the whole answer. What about scurvy? Lime juice solved the problem, but why? None of the known components of lime juice could have such an effect. That sour liquid must contain something unknown in very small, "trace" quantities.

Hopkins and Casimir Funk (1884–1967), from Poland, suggested that scurvy, along with several other diseases including beriberi, rickets and pellagra, were caused by diet deficiencies, the absence of what they called "accessory food factors," or "vitamines," as Funk named them in 1912. His term was later transformed to *vitamin.*

The first 30 years of the century saw amazing progress achieved against disease based on this "vitamin hypothesis," and in 1915, Joseph Goldberger (1874–1929), an Austrian-American physician, put it to work in solving the problem of pellagra.

Goldberger, as a very bright young man, entered New York City College at the age of 16 to study engineering. But he became fascinated with medicine, switched his course of study and became a physician instead. After a couple of years of private practice, which he found a little dull, he took the competitive exams for the Marine Hospital Service, on which he achieved the highest score, and joined the Public Health Service as a "microbe hunter,"

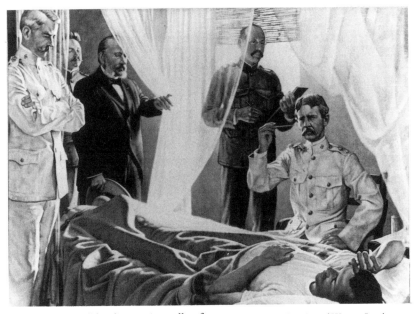

"Microbe hunters" battling against yellow fever   (Parke-Davis, Division of Warner-Lambert Company)

NEW FOCUS ON THE VERY SMALL:
THE ELECTRON MICROSCOPE

In the 17th century, Antony van Leeuwenhoek, a draper from Delft, Holland, amazed the scientific greats of the Royal Society in London with his accounts of creatures—"animalcules," he called them—so tiny that no one had ever seen them with the unaided eye. Using rudimentary microscopes that he had devised for examining textile threads, he thought of looking through them at droplets of water and plaque from his teeth, and, to everyone's great surprise, the world he saw teemed with minuscule creatures. From that time on, the world of the very small has provided rich territory for exploration by life scientists, and as tools have improved for observing tiny structures, knowledge has expanded dramatically.

In the 1930s, a new kind of microscope came into use for the first time that proved to have tremendous potential for probing the mysteries of living things: the electron microscope. Unlike van Leeuwenhoek's microscopes, and the many improved optical microscopes that followed, the electron microscope trains a stream of electrons instead of light rays on an object. These electrons are guided toward the specimen by magnets that deflect the electron stream, acting the part that glass lenses play in an optical microscope. Finally, in the type known as a transmission electron microscope, the beam passes through the specimen—much as X rays pass

battling yellow fever, dengue fever, typhus, typhoid and other infectious diseases all over the United States. In 1914, he was called in to solve the problem of pellagra, a disease that had plagued poor people in the American South for two centuries, but had recently become epidemic. An unpleasant disease, pellagra made the skin swollen, crusted and red, caused diarrhea and, finally, insanity. Everyone assumed that some infectious germ was the cause, but no one had ever succeeded in discovering it. Sometimes children, especially in orphanages, just seemed to "get over it" as they grew up.

Goldberger's approach to the problem was to observe. Without setting up laboratories or bringing in a microscope, he watched and listened. Pellagra was especially rampant in orphanages and asylums, but the staff never seemed to come down with it. If it was infectious, Goldberger reasoned, why didn't members of the staff catch it too? In his diary he wrote, "Although the nurses and attendants may apparently receive the same food, there is nevertheless a difference in that the nurses have the privilege— which they exercise—of selecting the best and greatest variety for themselves." Pellagra, he concluded, was a matter of diet.

through soft tissue—and is projected onto a viewing screen or photographic plate. Denser areas of a specimen let fewer electrons through; less dense areas let more through. In another type, the scanning electron microscope, the beam reflects off the specimen instead.

While the best optical microscopes could magnify as much as 2,000 times actual size, an electron microscope in the 1940s and 1950s could increase magnification as much as 250,000 to 300,000 times. In short order, biologists thought of using the new electron microscope on that curious creature the bacteriophage (a bacteria-devouring virus that revealed interesting insights into the physiology of both organisms), and in 1940 several workers in Germany published the first electron study of these viruses. Their work showed that infecting particles attached to the outside of the bacterial cell wall, an amazing bit of news about this creature's mode of operation. In 1942 Salvador Edward Luria, an Italian-American microbiologist, took the first good electron photomicrographs of a bacteriophage. And in 1945 Belgian-American cytologist Albert Claude (1898–1983) pioneered the use of the electron microscope in the study of cell structure. In that year he published the first detailed view of cell anatomy, revealing structures on a much finer scale than could have been imagined possible only 10 years earlier.

In the years to come, the electron microscope would come to play a more and more important part in discovering the nature of cells and their parts.

Then why the epidemic? After interviewing scores of doctors, Goldberger realized that hard economic times had made the local diet in many parts of the South even more meager than usual. From his research Goldberger was able to show that pellagra was caused by the absence of a vitamin in the diet. Most field hands were living on bread and molasses, a diet having no meat or any other good source of what was later discovered, in 1937, to be the missing factor: niacin, or vitamin B_2. The orphanage childen who recovered spontaneously were those who had grown up enough to begin to work, and meat had been added to their diets. Goldberger later did experiments inducing pellagra in dogs to make the discovery in 1923 that pellagra could be prevented with doses of brewer's yeast.

Ultimately, vitamin deficiencies were found to be the cause of beriberi (vitamin B_1), pellagra (vitamin B_2), scurvy (vitamin C), rickets (vitamin D) and certain problems with vision and night blindness (vitamin A). As a result, by the 1940s all of these ceased to be the major medical problems they had been in the past.

Joseph Goldberger, in search of the cause of pellagra   (Parke-Davis, Division of Warner-Lambert Company)

Overall, the period from 1895 to 1945 was a time when microbiology and biochemistry, building on the foundations laid by Pasteur, Koch and others, flourished dramatically, revealing countless secrets about how life functioned and what its processes were, as well as affording numerous dramatic breakthroughs in the medical sciences. These approaches would also come into play in the study of heredity, as the stories of the next chapter show—and the fields of biochemistry and microbiology, which grew so spectacularly during the first half of the century, still remain at the cutting edge of research in the life sciences today.

C H A P T E R 7

PURSUING THE TRAILS OF GENETICS AND HEREDITY

*E*arly 20th-century life science inherited two great ideas from the 19th century: evolution and genetics. The first was spearheaded by Charles Darwin's *Origin of Species*, which sparked great interest and considerable controversy when it was published in 1859. It naturally also raised the questions, How are traits passed on from generation to generation? What is the mechanism by which heredity operates? The 20th century would see a tremendous surge of interest in every aspect of these questions. Meanwhile, some of the answers were slowly emerging in an Augustinian monastery in Austria. This work, done by a monk named Gregor Mendel, was almost entirely overlooked in its time until it was rediscovered in the early 20th century.

GREGOR MENDEL

As a peasant boy Gregor Johann Mendel (1822–84) tended orchard trees for the lord of the manor; this occupation may have given him his start in botany. He later tutored to earn a living and finally in 1843 he entered an Augustinian monastery in Brünn, Moravia at the age of 21. Since the Augustinians supplied teachers to the schools in Mendel's native Austria, they sent him to the University of Vienna in 1851 to train in mathematics and science. Mendel apparently was a sensitive young man; he failed the examinations three times and had a nervous breakdown, but he finally completed the course and became a teacher in 1854.

Shortly after his university training, Mendel threw himself into a project that he always saw as an avocation, but which he pursued with the same passion for detail and experiment that any other scientist would. He brought

Gregor Mendel
(National Library of
Medicine)

to his experiments an unusual combination of talents in plant breeding and mathematics—which no one else at the time thought were related at all. And he embarked upon an amazing project, one that absorbed tremendous energy for years, for the same reason most scientists do: unyielding curiosity.

It all started like this: Soon after entering the monastery, Mendel had begun trying to breed different colors in flowers. This kind of enterprise wasn't unusual; plant and animal breeders had been controlling the outcome of breeding for centuries. In the process, Mendel acquired experience in the process of artificial fertilization in plants, and he also noticed something odd about his results. When he crossed certain species, as a rule, he would get the same hybrid results. But when he crossed his hybrid plants (whose parents had contrasting traits), sometimes their offspring had very odd traits. This puzzled Mendel, and he decided to try to find an explanation.

Mendel was not the first to notice this, of course. But no one ever had counted the number of offspring exhibiting the different forms or had tried to classify them. No one had tried to keep track of generations nor to do a statistical study. But to Mendel's mathematical mind, this seemed like a logical approach.

He began to develop a plan. He realized that he would have to raise many generations of many plants to obtain the kind of statistical information he needed. Otherwise, using just a few plants, he wouldn't have a large enough sampling and he might get misleading results. As he later explained in the introduction to one of his papers, "It indeed required some courage to undertake such far-reaching labors. It appears, however, to be the only way in which we can finally reach the solution of a problem which is of great importance in the evolution of organic forms."

From 1856 to 1864, he grew peas in the monastery garden, carefully keeping records of the traits from generation to generation. There, this unassuming Augustinian monk became the first person to formulate the cardinal principles of heredity. He conducted meticulous experiments in hybridization, carefully examining and recording details about thousands of plants—his best-known observations being made with garden peas.

He used peas because, over time, gardeners had bred pure strains: dwarf peas that always bred dwarf plants, and tall peas that always bred tall plants, for example. And peas were self-fertilizing but also could be cross-fertilized. This allowed for some interesting experimentation. He chose seven pairs of traits to observe that were easily identifiable and sharply contrasted, such as tall plants and dwarf plants, smooth seeds and wrinkled seeds, green cotyledons (the first embryonic plant leaf to appear from a sprouting seed) and yellow cotyledons, inflated pods and constricted pods, yellow pods and green pods, and so on. Mendel crossed a plant having one of these traits with a plant having the contrasting trait. To do this, he first removed the stamens from a flower to prevent self-fertilization. He then placed on the stigma of the flower a small amount of pollen from a plant with the contrasting trait. Next he wrapped the flower to prevent any further fertilization via wind or insects. The cross- fertilized plant would bear seeds, which he collected, catalogued and replanted to observe what traits would appear. Later he tried cross-fertilizing these hybrids with each other to see what would happen, always keeping careful records and noting what traits the offspring had. He repeated his experiments many times.

He found that if he crossed pure tall plants with pure dwarf plants, the hybrids that resulted were all tall. They looked just like plants produced from two tall plants. It didn't matter whether tall or dwarf plants furnished the male or female germ; the results appeared to be the same. Mendel called the trait that showed, in this case *tall*, the "dominant" trait; the trait that didn't show in this first hybrid generation, in this case *short*, he called "recessive." Next he crossed two of the hybrid plants together (both of which looked tall, but had a dwarf parent). He did several hundred of these crosses and found that he got some dwarf plants and some tall plants as a result. He counted them and worked out the ratio. There were 787 tall plants and 277 short ones—roughly three times as many tall plants as dwarf plants (3:1).

His results worked out as shown on the following chart:

	tall	short
tall	tall-tall	tall-short
short	tall-short	short-short

Mendel found the identical statistical distribution (give or take a few insignificant percentage points) for all seven traits he studied. He had, of course, focused on simple traits that had only two alternative forms. But because he had done that, he was readily able to perceive the pattern produced when he traced how parental traits were passed to their offspring. What emerged was the recognition that, while individuals exhibit many differences on the surface, beneath the surface even more complex differences existed. "It can be seen," Mendel later concluded, "how rash it may be to draw from the external resemblances conclusions as to their internal nature." Rash, at least, without careful controls, large numbers of offspring, statistical analysis and a big dose of caution.

Mendel didn't stop with the first generation of hybrids (those whose parents differed with respect to the tested trait); he continued some experiments as far as five or six generations. Later generations produced different, but always consistent, ratios. He also tried testing more than one trait at once, and as a result of his extensive experiments, he came up with conclusions that have since become known as the two Mendelian principles: the principle of segregation and the principle of independent assortment.

According to the principle of segregation, in sexually reproducing organisms (including plants), two units of heredity control each trait. When the reproductive cells are formed, though, the two units become separated (segregated) from each other, so that the offspring gets one unit for each trait from each parent. Mendel's work gave the first indication that inheritance might be carried by discrete particles in this way (which it is), and not blended.

When reproductive cells are formed, according to the principle of independent assortment, the distribution of the units of heredity for each trait doesn't interfere with the distribution of others. For example, he found that he could produce either tall or short pea plants that bore wrinkled peas and either short or tall pea plants with smooth peas.

Of course, we have to admit that Mendel came up with this clear, uncluttered picture as a matter of luck. The particular traits he chose in the pea plant had distinct, discontinuous variations for each pair, with no intermediate grades. And each was simple, not controlled by more than one hereditary unit. So he didn't end up, for example, with any offspring pea plants that were of medium height instead of tall or short. In humans, for example, we now know that height is controlled by several genes. So is skin color. So it's possible to get intermediate height and shades of skin color. Albinism (the absence of skin pigment) is a simple two-gene trait in humans, however, like Mendel's models. So a father and mother who have normal skin color but each carries a gene for the recessive trait of albinism could have a child that had no skin pigmentation at all. (In fact, the chances would be one in four.)

In addition, Mendel happened to choose traits that were not linked in any way—and he thought of the units of heredity as separate particles, which we now know they are not. So in his experiments, none of the odd results caused by what is now known as "linkage" occurred in his studies. (More on that subject later.) So, while his principle of independent assortment still applies, it holds only for those traits that are not linked.

Mendel had discerned patterns that no one else had ever seen, and his work should have made a big splash. But when he read his paper on the results of his experiments to the local natural history society, he met with complete silence and disinterest. No one felt moved to ask any questions. No discussion, animated or otherwise, ensued. Disappointed but not defeated, Mendel recognized that he was a completely unknown amateur, and so he thought he'd try to gain the sponsorship of a well-known botanist to back him. He sent his paper to the Swiss botanist Karl Wilhelm von Nägeli in the early 1860s. But the paper was too mathematical for Nägeli. To his credit, Nägeli did suspect that evolution came about in jumps, not in a smoothly continuous process. But Mendel's completely nonspeculative paper did not spark his interest, and he dismissed it with contempt. Mendel did succeed in publishing two papers in the obscure *Transactions of the Brünn Natural History Society* in 1865 and 1869, but few people noticed them. For those who did, they probably contained too much mathematics for the botanists and too much botany for those at home in mathematics.

And so, some of the most compelling information in the history of genetic science sat gathering dust for 35 years. Mendel died in 1884 without any idea that one day he would become famous and respected for his work.

MENDEL REDISCOVERED

Between the time Mendel did his work and the beginning of the 20th century, two important advances—improved microscopes and improved cell staining—opened up the nucleus of the cell to inspection and made possible an examination of hereditary factors at a different level. As a result, scientists began examining the cell nucleus and discovered that clusters of rods appeared shortly before cell division. These rods, or chromosomes, as they were named, split in two lengthwise, and the two sets that resulted went separate ways into the two halves of the dividing cell. Then the chromosomes rolled up into a ball and seemed to disappear. But no one knew what role chromosomes played.

In 1900 a Dutch botanist named Hugo de Vries had recognized that Darwin had not explained how individuals might vary and pass these variances on. So he began work on a theory about how different characteristics might vary independently of each other and recombine in many different combinations. He came to the conclusion from his study of the evening primrose (*Oenothera lamarckiana*) that new traits, or mutations, can appear suddenly and can be inherited, having found certain types of this plant that seemed substantially different from the original wild plant. Through experimentation he found that these substantially different forms also bred true thereafter. Before publishing his findings, de Vries looked back through the literature already published on his subject, and to his amazement he stumbled across the papers Mendel had published in the 1860s.

Coincidentally, two other scientists, one in Austria and one in Germany, also came across Mendel's work almost simultaneously. To the credit of all three, not one of them tried to claim Mendel's work as his own. All three published Mendel's results, giving him full credit, and added his own only as confirmation.

Some of the variations observed by de Vries were not actually mutations, as he thought, but hybrid combinations. Still the evidence for the theory from other sources reinforced his concept of mutations. Actually, mutations had long been commonly observed and used in breeding by those who raised cattle, sheep and other domestic stock. But, unfortunately, little exchange went on in those days between the scientific community and herders. So when Hugo de Vries published his book, *Mutationslehre*, in 1901, advancing the belief that evolution is due to sudden jumps or mutations (from a Latin word meaning "to change"), it was news. And he is commonly credited with the initiation of this line of investigation into the causes of evolution.

BATESON AND GENE LINKAGE

When the English biologist William Bateson (1861–1926) read the papers by Mendel that de Vries had found, he was impressed. He became a staunch supporter of the Mendelian legacy and translated the papers into English. And in 1905 he also took Mendel's work a step further based on his own experiments with inherited characteristics. He found that not all characteristics are inherited independently—some are linked, or inherited together—and he published his results in 1905.

By now the study of the mechanism of inheritance was gaining stature and a substantial literature, and Bateson proposed the name *genetics* for the growing new field. He was the first person anywhere to be appointed to the position of professor of genetics, which he accepted at Cambridge University in 1908. But it would be the American scientist Thomas Hunt Morgan who would explain how gene linkage worked.

MORGAN'S FRUIT FLIES

Few studies in genetics have been as fruitful as Thomas Hunt Morgan's work with the fruit fly (*Drosophila*), and one of his greatest inspirations was to use this small, easily bred creature to test his ideas about inherited characteristics and how they are passed on from generation to generation.

Descended from an illustrious Southern family (nephew of a Confederate general and related to Francis Scott Key, the author of the U.S. national anthem), Morgan joined the faculty of Bryn Mawr College after receiving

Thomas Hunt Morgan (National Library of Medicine)

his Ph.D. from Johns Hopkins University in 1890. By 1904 he became professor of experimental zoology at Columbia University.

Mendel's work had just come to light a couple of years before. Those who had observed the behavior of chromosomes in dividing cells and egg formation were talking about how close the fit was between this process and the results Mendel had published. But the human cell had only 23 pairs of chromosomes. They couldn't account for the thousands of characteristics in the human form, for example, unless there were some smaller structure, carrying some large number of factors, at work within the chromosome. In 1909 these factors were given the name *genes* (from a Greek word meaning "to give birth to"). No one at this time knew exactly what a gene was, though, or what mechanism it used to carry genetic information.

Ironically, as late as 1908, Morgan agreed with skeptics of Mendel's work, one of whom remarked, "There is no definite proof of Mendelism applying to any living form at present." But Morgan became curious about the process of mutation, and he looked around for a good species to study, just as Mendel had found the garden pea. He settled finally on the *Drosophila*, a small fruit fly that breeds with great rapidity, has clearly marked mutations, has only four pairs of chromosomes, and is easy to feed on a mash of bananas. Morgan found he could breed 30 generations in a year. Soon Morgan's room at Columbia University was filled with jars of flies.

However, despite close daily examination, Morgan couldn't find any mutations. He subjected the flies to high temperatures and to low temperatures. He exposed them to acids, alkalis and radioactivity. He fed them unusual diets. Still no mutants. Then one day in April 1910, after a year of watching and waiting, his wife spotted a fly with abnormal white eyes: *Drosophila* normally has red eyes. This was the mutant he had been waiting for. He bred his white-eyed male with normal red-eyed females. Soon he had 1,237 offspring. Every one had red eyes. However, among the following generation of 4,252 flies, 798 had white eyes. He had succeeded in perpetuating a mutation!

But there were two odd things about these figures. First, the ratio was not a Mendelian ratio of 3:1. What's more, all of the white-eyed flies were male. When Morgan and his team examined this question further, they discovered that the trait of white eyes was sex-linked: They had found the first linked traits, a complication that Mendel's clearly segregated traits hadn't raised.

Further study revealed that the chromosomes were the site at which the genetic material was located, and that genes were carried in linear fashion, strung like beads on a string or links in a chain.

By the end of 1910, Morgan had 40 different kinds of mutants, which he gave names like Humpy, Chubby, Pink-eye, Crumpled, Dumpy and Speck. Some had no wings. Some had no eyes. Some had no hair. Some had crumpled wings. And so on. For the most part, the mutations were not

This spectacular mutation in the fruit fly (Drosophila), *producing four instead of two wings (top illustration), compared to a normal fruit fly (below), is the result of recent research in genetics at California Institute of Technology.* (California Institute of Technology)

advantageous. But he found other links. White eyes developed only with yellow wings, never with gray wings. A characteristic he called ebony body went only with pink eyes, and black body went only with yellow wings. Morgan began to realize that certain characteristics were grouped together on the same chromosome.

Then one day, strangely, white eyes showed up in flies that did *not* have yellow wings. This was a real puzzler. Morgan boldly hypothesized that maybe a chromosome had broken and the broken piece had crossed over to reassemble with another chromosome. If this were so, a whole group of traits could be expected to be linked to another group of traits that they had never before typically occurred with. And that's exactly what proved to be the case. (This occurence became known as crossing- over.)

Morgan's work style typified the teamwork that became more and more prevalent in 20th-century science, incorporating the specialized talents of several scientists, including Alfred Sturtevant, a specialist in mathematical

THE PATHS TO DNA

No one in 1895 knew the significance of the three letters, DNA, that have since become almost synonymous with the field of genetics. Much less did anyone suspect that the powdery white substance called nucleic acid was not only essential to the hereditary process but carried the genetic code that transmits hereditary patterns from parents to offspring. But with the work already done by Gregor Mendel and rediscovered in 1900, scientists began to explore several paths of inquiry that led to the exciting burst of discoveries in this field that would come in the second half of the 20th century.

Out of the experimental work on mutations done by Morgan and his crew of talented researchers came the recognition that chromosomes carried the "factors" Mendel talked about (renamed *genes* in 1909 by Danish botanist Wilhelm Johannsen) and that these factors occurred in a line on the chromosomes, much like beads on a string. The famous "fly room" at Columbia also gave birth to the first recognition of crossovers during the formation of germ cells and the first successful gene mapping, the location of specific genes controlling specific traits along the length of a chromosome.

Meanwhile, Phoebus Levene (1869–1940), a highly original scientist who had fled from Russia to New York with his family in the 1890s to avoid anti-Semitism, began working on nucleic acids. By 1909, he showed that the five-carbon sugar, ribose, is found in some nucleic acids (which became known as ribonucleic acid, or RNA). In 1929 he showed that a sugar unknown up to that time, deoxyribose (which means, ribose minus one oxygen atom), could be found in other nucleic acids. All nucleic acids, it seemed, were divided into two groups, ribonucleic acid (RNA) and deoxyribonucleic acid (DNA). No one, however, had any idea what the function of these nucleic acids was. Few suspected them to be the genetic mechanism because they were relatively simple molecules, and genetic code, clearly, had to be complex. All one had to do was look around any natural setting—an open field or a wooded hillside or a lake—to observe the tremendous diversity produced by genetics. How could all this be directed by a molecule as simple as nucleic acids seemed to be? Most researchers assumed that the more complex protein molecules were much more likely candidates to contain the gene's secret mechanisms.

Levene also deduced the formulas of nucleotides and worked out the manner in which the components of the nucleic acids were combined into nucleotides (which are the building blocks of nucleic acid's very large molecule), and he hypothesized how nucleotides combined into chains.

But Levene's discoveries offer a good example of how research that seems quietly interesting, but not central, can suddenly take on new meaning as understanding unfolds.

In 1941, two American researchers, geneticist George Wells Beadle (1903–89) and biochemist Edward Lawrie Tatum (1909–75), were experimenting with mutations of a bread mold known as *Neurospora crassa*—an even simpler organism than Morgan's fruit flies. They found that some of the mutations with which they worked had lost the ability to form a substance necessary to their growth. From their studies of these and other mutations, they came to the conclusion that a gene's purpose was to supervise the formation of a particular enzyme—that is, that genes do their work by regulating specific chemical processes. When a mutation took place, a gene was altered so that it could not form a normal enzyme, and the regular sequence of chemical reactions that usually took place was interrupted, sometimes resulting in a radical change in the physical appearance or characteristics of the organism. Now it began to be evident that the experiments done by Morgan and others had tracked the visible evidence of changes produced by alterations at the biochemical level—in an organism's enzymes. Beadle and Tatum came to the conclusion that each gene supervises the production of just one enzyme, and their idea became known as the "one gene–one enzyme hypothesis." It was a major step toward understanding just what a gene was and how it worked.

Then in 1944, two Canadian scientists—Oswald T. Avery and Colin M. McLeod [mik LOWD]—working with an American bacteriologist, Maclyn McCarty, located the carrier of genetic information in cells: the DNA molecule. Working with the pneumococcus bacteria, they experimented with two strains, a smooth (encapsulated) strain and another strain that had a rough (nonencapsulated) exterior. They removed the DNA from a smooth strain and added it to a strain of rough bacteria cells. To their amazement, the offspring of the rough cells were smooth! Somehow, the DNA from the smooth strain had transformed the genetic code of the rough cells so that the characteristics passed on were not the rough exterior of the parents, but the smooth exterior of the cells from which the DNA had been taken.

Immediately, DNA took center stage and stayed there. How could the simple DNA molecule contain the intricate and complex information necessary for even such a straightforward genetic message as smooth exterior versus rough exterior? What could its structure possibly be? What was the role of that other nucleic acid, RNA? The stampede was on to find the answers. By 1953, Francis Crick and James Watson would make their famous breakthrough model of the DNA structure, but many questions remained. Today, the search continues for ever more detailed answers about the molecular basis of heredity and the mode of transmission and transcription of genetic code—the instructions that guide offspring in repeating the biological patterns of their ancestors.

analysis of the results achieved through breeding the *Drosophila* and mapping of genetic factors on chromosomes; Hermann Muller, a theorist who also had a talent for designing experiments; and Calvin Bridges, who was especially adept at studying cells. They worked both independently and as a unit, sharing results and collaborating on experiments. Together they worked out the idea that Mendel's factors, as he had called them, were specific physical units, or genes, with definite locations on the chromosome.

With Sturtevant, Muller and Bridges, Morgan published *The Mechanism of Mendelian Heredity* in 1915, providing an analysis and synthesis of Mendelian inheritance as formulated from the author team's epoch-making investigations. A classic work, the book has become a cornerstone of modern interpretation of heredity.

With his book *The Theory of the Gene*, published in 1926, Morgan established gene theory and extended and completed Mendel's work as far as it could be taken with the tools available—brain, eye and microscope. A student of Morgan's would use the X ray to probe the genetic code, but no further major progress would be made in genetic investigation until the advent of molecular biology and the work of Francis Crick and James Watson a generation later. Morgan won the Nobel prize in physiology or medicine in 1933.

DARWIN AND MENDEL COMBINED

Ever since Hugo de Vries had discovered the work of Gregor Mendel, the Augustinian monk's extensive experimental genetic evidence and Darwin's evolutionary theory still had not been fused successfully. In general, geneticists had assumed that there were normal genes and occasional mutations. Most of the latter would be weeded out, they thought, but the few useful mutations would produce evolutionary change. That's where Russian-American geneticist Theodosius Dobzhansky [dub ZHAHN skee] entered the scene.

Dobzhansky (1900–75) was born in the Ukraine and educated at the University of Kiev, during the chaos of the Russian Revolution and civil war, graduating in 1921. He taught at Kiev and Leningrad and then left Russia to work at Columbia University in New York City with T. H. Morgan. He also accompanied Morgan to the California Institute of Technology, where he joined the faculty. He became a U.S. citizen in 1937.

In his book *Genetics and the Origin of Species* (1937), Dobzhansky successfully combined the ideas of Darwin and Mendel. Mutations, he showed, are in fact very common and are often both viable and useful. So Dobzhansky threw away the concept of a "normal" gene. There are only "those that have

Theodosius Dobzhansky (National Library of Medicine)

survived," and those that do not. Which genes survive depends on which are encouraged by chance and local conditions at the time. What works for one habitat may not for another. What's successful and useful at one time may be fatal at another time.

The considerable work done by de Vries, Bateson, Morgan, Dobzhansky and their colleagues on genetics in the first half of the 20th century went a long way toward answering questions about the genetic process. And by 1945, geneticists were rapidly unraveling the mysteries of the microscopic world that governs the shape of all living things.

IN SEARCH OF ANCIENT
HUMANS

As Charles Darwin's theory of evolution gained wider, but by no means universal, acceptance during the later days of the 19th century, many scientists began to search for fossils that they hoped would either prove or disprove Darwin's theory of human evolution. For the "evolutionists," the search centered primarily on the hopes of finding the so-called missing link, a fossil that many believed would graphically establish humankind's relationship with its apelike ancestors. Contrary to what many incorrectly believed Darwin's theory to say, he had not claimed that humans had descended directly from the apes, but that both humans and apes were related through a common ancestor. So the fossil hunters postulated that if humankind and the apes shared a common ancestor it might be possible to discover the fossil remains of some extinct intermediary form, linking humans and apes, that would have some characteristics of both.

For the much smaller and less active group of "antievolutionists," led by Darwin's nemesis, the zoologist and paleontologist Richard Owen (1804–92) among others, the search was for very ancient human fossils, which they believed would show that humans had not gone through any evolutionary process, but instead had always been "completely human" and had changed very little or not at all throughout time. In short, although most antievolutionists had accepted the idea that the Earth was very old, they believed that humans had started out "human" long ago, so that even very ancient fossils of our ancestors, if they could be found, would look much like their modern descendants.

As to the so-called similarities between humans and apes, Owen and others also had strong objections. Although there are some structural similarities between humans and apes, Owen contended, the major differences are far more numerous. He was most concerned with those differences that were not subject to external and environmental influences and so would be passed down through many generations without modifications. One of his favorite arguments had to do with the pronounced eyebrow ridge of the gorilla. Since there is no muscle attached to the ridge, Owen argued, and since there was nothing in the gorilla's behavior that implied the ridge could be altered by external causes operating on successive generations, therefore the ridge should occur in all the gorilla's ancestors as well as all of its descendants. Thus, Owen argued, if humans and gorillas shared a common ancestor, then humans should also have such a prominent eyebrow ridge, which rarely ever actually appeared in humans. It was obvious, Owen contended, that humans and apes could not share a common ancestry.

Ironically, just such an eyebrow ridge was a major and disturbing feature of one of the first fossil finds eventually to be accepted as a fossil of an early human.

THE NEANDERTHALS

In 1856, quarrymen working in a limestone cave near the village of Neander, Germany accidentally unearthed Europe's first known human fossil. The remains consisted of a heavy skull cap and more than a dozen bones. When the fossils came to the attention of a local schoolteacher, he had them gathered up and sent to an anatomy professor, D. Schaaffhausen, in Bonn. After studying the intriguing bones, Professor Schaaffhausen presented his conclusions at a meeting of the Lower Rhine Medical and Natural History Society in Bonn in 1857. The bones, he said, were human and very ancient, but they were the remains of a being unlike any humans presently known in Germany. The limb bones were very thick and misshapen with pronounced muscle attachments, and the strangely shaped skull had heavy eyebrow ridges "characteristic of the facial conformation of the large apes." Professor Schaaffhausen's conclusion was that the bones must have belonged to an ancient and ferociously barbaric northern tribe that had probably been conquered by the Germans very long ago.

As news of the find and the professor's conclusions spread, though, other scientists were quickly heard from. By 1861 Neanderthal Man, as the disturbing bones had come to be called, was at the center of intense controversy. Who or what exactly was Neanderthal Man? Was he, as the evolutionists believed, a very ancient early stage of human development? With his brutish, apelike brow and thick, bent bones, did he in fact offer dramatic evidence that linked humans to some apelike ancestor? Or was he, as the antievolutionists believed, simply a monstrously deformed modern human, very likely an extremely contorted and unpleasant-looking idiot who had probably lived as a hermit shunned by society and died in the cave in which he had been found? Since no other fossils, such as fauna, had been found in the cave that might have helped to establish Neanderthal's age, and today's sophisticated dating techniques had not yet been developed, the antievolutionists' argument appealed to many who watched the controversy from the sidelines. Certainly, given its appearance and all that appearance seemed to imply, there was something uncomfortable in believing that such a creature could in any way be related to modern humans. Even some evolutionists were troubled when confronted with Neanderthal's disturbingly brutish appearance. One prominent evolutionist, William King, a professor of geology at Queen's College, Galway, in Ireland, was so disturbed by Neanderthal's appearance—which suggested to him a being with "thoughts and desires . . . which never soared beyond those of the brute"— that he proposed Neanderthal be assigned its own species: *Homo neanderthalensis*, which would remove it altogether from association with *Homo sapiens*, the name given to the human species.

Was Neanderthal the evolutionists' missing link? The majority conclusion among the evolutionists at the time was summed up best by "Darwin's Bulldog," the brilliant and pugnacious evolutionist T. H. Huxley. Huxley argued that although the skull was the most apelike that had yet been discovered, it did not come from a being who was intermediate between ape and man. The determining factor, Huxley argued, was the size of the brain. Neanderthal's cranial capacity was almost twice the size of the largest apes, and well within the range of modern humans. Neanderthal was very ancient, disturbingly brutish—and human. But it was not the missing link. Neanderthal was indeed primitive, to many people frighteningly so—but it was not primitive enough.

In the following years many more fossils, both male and female, of these strange humans with heavy jaws and eyebrow ridges began turning up around the world. Cartoons in the popular press quickly shaped the image of the brutish, shambling "ape man" that later became a fixture in low-budget Hollywood motion pictures, and even today mystery surrounds their existence. We know now that the first Neanderthal fossil was not completely typical of the later finds, that much of the twisted deformity of the bones was probably due to an arthritic disease, and as one anthropologist has suggested, with modern clothes on and hats pulled down over their heads, Neanderthals could probably walk down a busy street today without attracting more than slight notice.

How, though, is Neanderthal related to modern humans? Modern estimates indicate that Neanderthals lived between 100,000 and 40,000 years ago. For many years investigators believed that Neanderthals were on our direct ancestral line, but more recently some surprising fossil discoveries have shown that Neanderthals and *Homo sapiens* (modern humans) apparently existed at the same time. What relationship then did they have with us—and how and why did Neanderthal disappear? Were they killed off by modern humans? Did they interbreed to be gradually absorbed into the modern gene pool? Or did they simply lose out in the battle for survival and gradually wither into extinction?

Neanderthal wasn't the missing link, but many scientists, including the always skeptical T. H. Huxley, were convinced that fossil evidence of a true link would be found. It was late Victorian optimism at its most exhilarating, given the scarcity of fossil evidence at the time. Fossilization is at best a hit-or-miss process that even today is not completely understood, but the effect is that in some circumstances a plant or an insect or a bone, instead of breaking down into its chemical components after death, becomes buried and gradually infiltrated by minerals within the soil, which slowly replace it molecule by molecule until the original organic form is replaced and duplicated in its exact form by stone. The key phrase here is "in some circumstances," and those circumstances are so rare that only an infinitesimally small number of bones become engaged in the fossilization process. Even today if all the bone fossils directly related to the story of human evolution were gathered in one place they would not fill up one small room.

By the late 19th century, while the vast majority of scientists had swung in favor of evolution, there were still some disagreements over how and when humans had diverged from their ancient ancestral stock. While there was general consensus that three primary attributes could be considered distinctly human—the enlargement of the brain, consistent upright walking, and the dental arrangement of small front teeth and larger back teeth—there was controversy about which attribute came first. It was a controversy that colored many of the early attempts to find the so-called missing link. For some 30 years after the discovery of Neanderthal Man, no new fossils were discovered that would help define exactly when humans had "become human," or which trait marked the way.

HOMO ERECTUS

Ernst Haeckel (1834–1919) was a German evolutionist and a popularizer of evolution who surpassed even the brilliant and eloquent T. H. Huxley in his appeal to mass audiences. But Haeckel was frequently also an embarrassment to many scientists because, although he was a charismatic speaker and visionary in his public lectures, his research and science were often sloppy and deeply flawed. Haeckel argued that the most important human attribute was the power of speech and that if the missing link were to be found it would be at that point in evolution just before humans acquired speech. In his book *The History of Creation*, published in 1868 and one of the earliest zoology texts to entirely embrace evolutionary theory, he also laid out the first ancestral trees, which we now see in nearly every biology or anthropology book. As Haeckel saw it the evolution of life was presented from a single cell through 22 stages of development up to modern humans, who stood exalted at the top of the tree. Scientists today, of course, see Haeckel's "tree" as highly misleading since it depicted a direct evolutionary path leading specifically to humans as the final goal of evolution. But in Haeckle's time his work inspired many would-be fossil hunters, particularly because at stage 21 there was a creature he said was the notorious intermediate link between man and apes. *Pithecanthropus alalus*, he called it, or "ape-man without speech." And, he suggested in his book and lectures, he even knew where to find it! Haeckel believed that *Pithecanthropus alalus* had originated on an ancient continent called Lemuria, later sunk beneath the Indian Ocean. From this "cradle of humanity," he said, his speechless ape- man spread first to Africa, Asia and Java, and then to other parts of the globe. Somewhere there, probably in southeast Asia, in Borneo or Java, he proclaimed, would be found the famous missing link.

Haeckel didn't go hunting for *Pithecanthropus alalus* himself; he preferred to inspire students and fossil hunters to do so. One who was just so inspired was a Dutch anatomy teacher, Eugène Dubois (1858–1940). Bored by his sedentary teaching life and obsessed with the idea of finding the missing link, Dubois announced to his startled colleagues that he was leaving his teaching post in Amsterdam and following Haeckel's advice, heading for Asia to discover the grand evolutionary prize. Finding no financial support for his quest, he enlisted in the Dutch East India Army, promising them eight years of his life but in turn getting himself assigned to Sumatra. With his wife and small daughter he set sail in the autumn of 1887. It was the turning point in the young evolutionist's life.

Given light medical duties in Sumatra, Dubois spent every minute of his spare time searching the area's limestone caves and deposits for his elusive quarry. Weakened by an attack of malaria, he managed to convince the Dutch East Indian government to relieve him of military duty altogether and instead assign him to head a paleontological survey in Java. The speed with which he managed to set up his expedition led some people to suspect that the government had been aware of his intentions all along and had in effect, by accepting his enlistment, furnishing his transportation and even providing convict labor, covertly agreed to finance his operation. In any case, by March 1890 he was on his way to Java, and his dreamed-of expedition to find the missing link was now completely under way.

He found his first fossil in 1891. Or at least, his convict laborers found it for him. Painstakingly combing a stratified embankment along the Solo River, his workers turned up a tiny jaw fragment and a molar tooth. Weeks later came a skull cap, and finally, almost 10 months later, a fossil thighbone

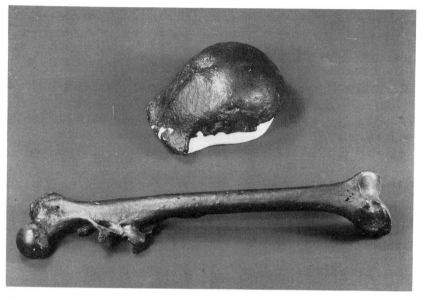

Casts of a skull cap and femur (thighbone) found by Eugene Dubois. He identified them as belonging to a hominid later named Homo erectus. (Neg. No. 319781, Courtesy Department Library Services, American Museum of Natural History)

from the same site. The tooth was chimpanzee-like, the skull cap had a small cranial capacity and thick browridges, and the thighbone, although heavier and thicker than that of a modern human, was definitely human and belonged to a being who obviously had a habitual upright stance. It was slim evidence, but Dubois pondered it. The skull cap was too large for an ape, although its brain capacity was small, and it had the browlike ridges. The small fragment of jaw didn't give enough evidence to determine whether the creature could have spoken or not, and so Dubois couldn't apply Haeckel's theory. But the thighbone spoke volumes. His fossils, he determined, indicated a small brained ape-man who walked upright on two legs.

He had, he decided, discovered the missing link! For, as he wrote in 1894, "this was the man-like animal which clearly forms such a link between man and his nearest known mammalian relatives as the theory of development supposes. . . ."

Since he was relying primarily on the thighbone for his evidence, not having enough jaw bone to support Haeckel's contention, he called his fossil *Pithecanthropus erectus* (upright ape-man), instead of Haeckel's *Pithecanthropus alalus* (speechless ape-man) and offered his evidence to the scientific world.

118

The members of the scientific world, taking the full case into account at the Third International Congress of Zoology, at Leiden in 1895, both liked and disliked what they saw. There was little disagreement that Dubois had made an amazing find, and one of great importance. Only a few other examples of fossil man had been discovered up to that time, and any addition to the growing study was of immense value. But few accepted Dubois's contention that he had indeed found an intermediate between man and ape. Some, disturbed by the poor records kept on the original finds and by the fact that exact locations and times of the finds had not been pinpointed, questioned whether the fossils really represented only one individual. Some felt that the fossils were more ape than man, while others argued that they were more man than ape. Some reasonably insisted that there was just not enough fossil evidence to make a decision one way or another.

As similar congratulations ("good find!") and objections ("bad interpretation!") greeted him at other meetings, Dubois became increasingly disappointed and angry. He was determined that his interpretation was correct, that he had found the missing link, and he felt that only jealousy or stupidity kept others from seeing it. Passionately possessive of his find, he lugged his Java man, as scientists began calling it, from meeting to meeting, seeking recognition and carrying it in a small suitcase as if it were a beloved pet. In the end, however, beaten by frustration and exhaustion, he withdrew from the scientific community, taking his fossils with him.

Dubois's story doesn't have a happy conclusion. For the remainder of his life he lived in virtual seclusion from the scientific community, refusing to let anyone again study or even see his beloved *Pithecanthropus*. According to one popular legend he even went so far as to hide the fossils beneath a floorboard in his house, taking them out occasionally for his own satisfaction. Finally, however, when a spectacular series of finds began turning up in China in the 1920s and 1930s, the true nature of Dubois's fossils came into focus.

The story of Peking man, as these China finds came to be called, was a curious one. In the early 1920s a German naturalist by the name of K. A. Heberer became aware that apothecary shops in Chinese cities and towns had been selling "dragon bones" to be ground up and used as folk medicine. Curious, he began visiting the shops and examining the bones. In the process of poking around in the bins from which the loose bones were sold, he discovered specimens from more than 90 fossil mammals and a fossilized tooth that appeared to have belonged either to a human or an ape.

Word of Heberer's discovery quickly spread through the large international community of bone hunters who had been swarming through China since Haeckel's "cradle of humanity" theory had become popular. The next few years saw a carnival of bone hunters scrambling through shops, caves and hillsides.

Finally, in 1929, after years of assorted finds, the first hominid skull turned up buried deep in a limestone cave near the village of Chou K'ou Tien. (A hominid is any member of the family of two-legged primates that includes humans, both extinct and living.) It was heavy-boned, with prominent browridges and a smaller-than-human brain capacity. The site proved to be a treasure trove, and throughout the early 1930s more than 14 skulls were turned up along with 11 jawbones and more than 100 teeth. There were some strange peculiarities: mysteriously, mostly heads and parts of heads were discovered at the site, and most of the skulls showed evidence of the brains having been scooped out. But all of the skulls bore a striking resemblance to Dubois's Java man. The resemblance was so striking in fact that there could be little doubt that Java man and Peking man both belonged to the same species—an extinct species of upright-walking humans with slightly-smaller-than-modern human brains, who lived 1.8 million years ago. In 1950 the evolutionary biologist Ernst Mayr named the species *Homo erectus*, defining it once and for all as not an ape or a missing link, but an early form of man.

Up until his death in 1940, Eugene Dubois, whose Java man is recognized today as the first discovery of *Homo erectus*, refused to accept the Peking finds as resembling his own. In his last days, the bitter Dubois even reversed his claim that his beloved *Pithecanthropus erectus* had even the slightest connection with man—claiming instead that it had nothing human about it and was an advanced form of fossil ape!

In his obituary of Dubois, the anatomist Sir Arthur Keith summed Dubois up best. "He was an idealist, his ideas being so firmly held that his mind tended to bend facts rather than alter his ideas to fit them."

THE PILTDOWN HOAX

It was an ironic statement coming from Sir Arthur Keith, who in the early 1920s had allowed his own preconceptions to deceive him into authenticating one of the most notorious of all scientific hoaxes.

Like many of his English colleagues, Keith belonged to the school of evolutionists who believed that development of a large brain, not upright walking, was the first primary trait to distinguish the human line. Reflecting on the Neanderthal and Java fossils, Keith once told an audience, " . . . we have knowledge—a very imperfect knowledge—of only two human individuals near the beginning of the Pleistocene period. The one was brutal in aspect, the other certainly low in intellect." If those fossils truly represented the ancestors of modern humans, he argued, then we had to accept the fact that " . . . in the early part of the Pleistocene, within a comparatively short space of time, the human brain developed at an astounding and almost incredible rate."

Keith believed instead that Neanderthal and Java man could not be modern man's true ancestors, but were instead cousins and contemporaries of modern man's large-brained ancestor.

It is little surprise then that when a stunning fossil find came to light from a gravel pit near Piltdown Common, in Sussex, England, Keith would become intimately involved with the notorious story of its identification.

The story began with an amateur English geologist named Charles Dawson. At a meeting of the Geological Society in December 1912, Dawson read a paper describing his activities and finds at Piltdown Common. A few years before, Dawson said, he had become familiar with some unusual brownish flints that he had been able to trace back to the gravel pit. Thinking that the spot might hold other specimens of interest, he had asked some laborers who sometimes worked the pit to keep their eyes open for anything unusual they might turn up.

What they turned up, a short while later, according to Dawson, appeared to be a part of a coconut shell that had been broken by their work. They had

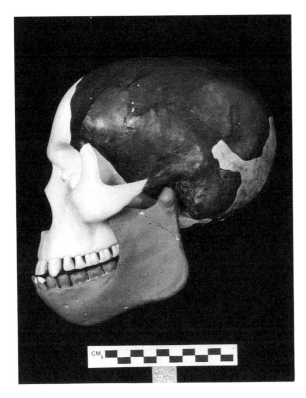

A reconstruction of the Piltdown skull, whose large cranium and ape-like jaw seemed to indicate that the emergence of human traits followed the development of a large brain. As it turned out, the Piltdown cranium had come from a human and the jaw had come from an ape—patched together to look like they came from the same creature. (The Natural History Museum, London)

121

thrown away the other pieces, but then, wondering if Dawson might be interested, had retrieved one small piece to show to him. Dawson said that he recognized at once that the fragment was a part of fossil skull. He returned to the pit to attempt to retrieve the rest of the pieces, but it wasn't until some time in the autumn of 1911 that he at last found another skull fragment.

At this point, Dawson said, he had taken his fragments to the British Museum, where he showed them to Arthur Smith Woodward, the museum's keeper of geology. Woodward had become interested and had begun to spend weekends working the pit with Dawson. The two men were shortly joined by the Jesuit priest and amateur geologist Father Pierre Teilhard de Chardin, who also began to put in some time at the dig.

In the summer of 1912 they found more pieces of the skull and an assortment of fossilized mammal teeth, and they made their biggest discovery: part of a fossilized apelike jawbone.

Their discovery, Dawson announced triumphantly at the crowded 1912 meeting, was a new species, *Eoanthropus dawsonii*: Dawson's Dawn man. Even more exciting was what Dawn man, or Piltdown man as it quickly became known, represented. With its parts reconstructed, Piltdown man could easily be seen to have been an extremely ancient large-brained individual with a strikingly apelike jaw.

It was what many English evolutionists hoped to find—particularly those who believed that it was the large brain that had first marked the emergence of distinctly human traits. Here, in Piltdown man, with its large-brained human cranium and its apelike jaw, was unquestionable proof. And the find had been made right in England, where, these English anthropologists thought, intelligence obviously must have developed early.

Sir Arthur Keith, who had some quibbles with the initial reconstruction, nevertheless was delighted. So were many other leading British anatomists and evolutionists, who quickly joined with Keith in examining Piltdown and declaring it authentic. A few voices murmured dissent, especially when it was claimed that the other fossil evidence unearthed in the pit established that the large-brained Piltdown must have predated both the Neanderthals and Java man; but they were quickly drowned out. So too were voices that suggested that both the human cranium and the apelike jaw were almost a little too perfect. The skull was just too human and the jaw too much that of an ape: How could the two ever really have been attached to the same individual?

The answer of course was that they couldn't be. Piltdown was a fake. The skull had come from a human and the jaw had come from an ape; both had been carefully worked on, broken, chipped and stained to appear as if they belonged together.

Unfortunately it took more than 40 years for this to be discovered—until the 1950s, when a young scientist, not locked into the views of his

A discussion on the Piltdown Skull

Many of the most respected British anthropologists of the time were fooled by the Piltdown "fossil"—their perceptions blurred by their own biases. Charles Dawson (third from left) and several eminent researchers look on as Arthur Keith (seated, center) examines the skull in this painting done in 1915 by John Cooke. (The Natural History Museum, London)

predecessors, applied unbiased examination and new scientific tests to the fossils. Keith and his colleagues had seen what they wanted to see and ignored the obvious.

Who had constructed the hoax? Dawson, one of his colleagues at the site, someone else? Many theories and accusations have been put forward, but so far the answer remains a mystery, and it is likely to continue so.

AFRICAN BONANZA

While Piltdown sat in its prestigious place in the British Museum waiting its final downfall, other startling discoveries were being made elsewhere in the world—important discoveries that, sadly, would be ridiculed for years by the misguided champions of Piltdown man.

In the first quarter-century, informed opinion generally went along with Haeckel's theory that the likely origins of the human family would be found somewhere in Asia. It was obvious from finds such as Neanderthal and others, including the controversial Piltdown, that Europe had also been an important area in the story of human evolution, but few bone hunters had bothered to follow up Charles Darwin's suggestion that Africa was actually the most likely place to discover the very earliest origins of humans.

In 1923, the 30-year-old Australian-born anatomist Raymond Dart arrived in South Africa to begin a teaching position at the University of Witwatersrand in Johannesburg. Dart had studied in England under some of the leading British evolutionists and had developed a passionate interest in evolution, particularly in the hunt for early humans. His important part in discovering the trail of human evolution began in 1924 when a student of his brought him a fossilized baboon skull that she had seen on the mantlepiece in a friend's living room. The home belonged to the director of a commercial quarrying firm, the Northern Lime Company, and the skull had come from a place called Taung about 200 miles away. Intrigued by the skull, which appeared different from any other baboon skulls that he had ever seen, Dart asked that word be given to the quarry workers to send him any other such finds they might have turned up. Within weeks he received two packed boxloads of fossils! Like a child on Christmas morning he began excitedly to investigate his treasures.

Painstakingly sorting through the boxes, he hit the jackpot! Around the world much of the discussion in evolutionist circles was about brains and brain sizes and which came first, upright walking or large brain, but no one had found a specimen anything like what Dart was suddenly holding in his hand: an endocranial cast—a lump of fossilized material that exactly copied the inside form of the skull, preserving details of the outer form of a brain, blood vessels and all. It wasn't a fossil brain, but it was the next best thing to it! It was big for an ape, and small for a human, but there were features that he recognized as being more human than ape.

His luck was only beginning. Excitedly continuing his search, he found a block of limestone in which the back of a fossilized forehead and a partial fossilized face were embedded. The endocast fit into an exact position in the limestone rock! It was part of the same single specimen! What, though, would the full face look like? For the next 73 days he carefully chipped away at the rock, bit by bit, slowly exposing the encrusted face, "No diamond cutter ever worked more lovingly or with such care on a priceless jewel . . . ," he would later write, describing his task. On December 23, 1924, his present opened up before him. "I could view the face from the front, although the right side was still imbedded. . . . what emerged was a baby's face, an infant with a full set of milk teeth and its permanent molars just in the process of erupting," he wrote. "I doubt if there was any parent prouder of his offspring than I was of my Taung baby on that Christmas."

The fossil was that of a child, probably three to four years of age at the time of its death. And it was, Dart was certain, a major find, some kind of early hominoid (the name of a group that includes humans, gibbons, and apes), but what kind? It was not likely to have been a forest ape, since South Africa had a relatively dry climate going back for millions of years. And, although the fossil did appear to have the face of a young ape, its brain was much more human—not in size, or the size it would have grown to in adulthood (it would still have been much too small to be completely human)—but in some certain characteristics of its architecture. More important, Dart observed that the aperture through which the spinal cord leaves the cranium and enters the spinal column was farther forward in the Taung child than it was in modern apes, indicating that the head had been balanced above the spine as it is in erect bipedal creatures. In addition, the slant of the forehead was not as pronounced as it is in apes. Dart thought long and hard about his findings, checked and double-checked his fossils and his thinking, and came to his conclusion. His Taung child represented a very ancient creature (scientists today estimate that the Taung child lived between one and two million years ago), who had a brain only a little larger than an ape's but a brain structure that in some aspects suggested some humanlike rather than apelike features. In addition, although the creature had the face of an ape, it walked habitually upright like a human.

On February 7, 1925 Dart published his news about his discovery and conclusions in the British scientific weekly *Nature*. Included in the now famous paper was Dart's suggestion that his Taung child was a member of an "extinct race of apes intermediate between living anthropoids and man."

Unlike Dubois, he hadn't gone in search of the missing link, and he was too sophisticated an evolutionist to believe that there was one single, clear-cut "link" that would answer all of humankind's questions about its past. But, he was certain that his find, which he called *Australopithecus africanus* (southern ape of Africa) represented an entirely new family never encountered before, a partial missing link perhaps, and certainly an important piece in the puzzle of human evolution.

His article raised an immediate and violent storm of disagreement. The weight of the scientific community fell down like a rock on his shoulders. Not only was Dart working in an area, Africa, which everyone knew was not the right one to be looking in—Asia after all was the promising "cradle of humanity" that would supply the proper fossils—but his interpretation of his find was wildly off the mark. At least so the "experts" loudly declaimed. Sir Arthur Keith, examining a plaster cast of the Taung child later that summer, summed up most expert opinion when he wrote to *Nature* challenging Dart's conclusion that Taung was a midway point between ape and human. "An examination of the cast," he wrote, "will satisfy zoologists that this claim is preposterous. The skull is that of a young anthropoid ape—one that was in its fourth year of growth, a child—and showing so many points of affinity with the two living African anthropoids, the gorilla and chimpanzee, that there cannot be a moment's hesitation in placing the fossil form in this living group." As for the matter of Dart's claim that the endocast indicated some humanlike features of the brain, Keith disclaimed that as "a matter of guesswork."

Besides, hadn't Piltdown man established once and for all that the early human brain was a large one and preceded upright walking? The upstart Dart had it all wrong! And all wrong in Africa of all places.

Thus, with a collective wave of the establishment hand, Dart and *Australopithecus africanus* were summarily dismissed.

Unlike Dubois, though, Dart was not about to bury his find or his ideas. Despite lack of funding or official endorsement, he continued his quest to discover more evidence, believing always that it eventually would be found and his Taung child vindicated.

This time the story had a happy ending.

"It makes one rub one's eyes. . . . Here was a man who had made one of the greatest discoveries in the world's history—a discovery that may yet rank in importance with Darwin's *Origin of the Species;* and English culture treats him as if he had been a naughty schoolboy. . . ." So wrote Robert Broom, a Scottish paleontologist and doctor who supplies the next chapter in Dart's story.

Variously described as eccentric, egomaniacal, uncouth, uncivilized, pushy, and pig-headed, Broom had already established his reputation as a South African fossil hunter when he decided to champion the cause of Dart and his Taung child. Born in Scotland in 1866, Broom had spent some time in Australia before beginning a medical practice in South Africa around 1900. Broom's bone-hunting specialty was the fossils of early mammallike reptiles in South Africa. He was determined, energetic and successful. In 1920 he was elected to the Royal Society, and in 1928 he received the society's Royal Medal for his mammal work. When Broom read Dart's paper in the February 1925 issue of *Nature* he immediately wrote to Dart and congratulated him on his "glorious" find. Two weeks later, unannounced, he burst into Dart's lab and fell on his knees in front of the bench that held the Taung skull. It was a typical dramatic action for the melodramatic Broom, and after a weekend spent examining the skull to his own satisfaction Broom committed himself to finding other fossils that would establish its authenticity.

Unfortunately, previous commitments kept Broom from devoting much time to his search for other australopithecines until, at the age of 69, he began a series of full-scale expeditions in 1936.

After finally getting started, he worked with a vengeance. He admired Dart, but "Dart was not much of a fighter," he wrote later. Broom was prepared to fight for both of them.

His results were astonishing.

Putting the word out that he was on the search, he learned from a couple of Dart's students that there was another large commercial limeworks at Sterkfontein near Johannesburg. He contacted the managers of the works and asked them to keep their eyes out while he continued investigating other sites. Checking in on the limeworks in August 1936, only a few months after beginning his investigations, Broom began rummaging through piles of collected debris that the managers thought might be interesting to him. Astonishingly, he quickly turned up a beautiful adult *Australopithecus* skull— and a completely intact adult endocranial cast!

His luck and his adventures continued. In 1938 he learned that a school-boy at Kromdraai, about a mile from Sterkfontein, was in possession of some unusual-looking teeth. Tracking the boy down in the middle of a school day, he invaded the schoolroom, bought the teeth on the spot and learned that the youngster had a stash of other treasures hidden away nearby. There

Robert Broom (Neg. No. 34215, Photo: Julius Kirschner, Courtesy Department Library Services, American Museum of Natural History)

was still an hour or so until school let out for the day, and not wanting the youngster to get out of his sight, Broom convinced the schoolmaster to let him lecture to the students. For the next hour, scribbling illustrations on the blackboard, he lectured on fossils and caves to four teachers and 120 students. As soon as the bell rang, he was off with the youngster to a nearby hill where the boy had hidden an *Australopithecus* jaw with two more teeth still attached.

He continued to unearth more sites and more *Australopithecus* fossils, more skulls, jaws, teeth, a nearly complete pelvis, sections of shoulderblade and arm and leg bones. In the late 1940s, Dart, who returned to the field after a few years' absence, also uncovered an important site at Makapansgat. By 1948, Dart and the 81-year-old Broom had collected enough fossil evidence to convince most of the once-skeptical scientific establishment that *Australopithecus* had existed, in not only one form, but two—one slender and the other more robust and stocky. Both forms, though, had possessed the smallish brain, apish face and upright walking habits of Dart's original find. And both forms were very ancient relatives of humans. Even the skeptical Arthur Keith was forced to reconsider his earlier evaluation in the light of the overwhelming evidence pouring out of Africa.

"Professor Dart was right and I was wrong," he wrote in a letter to *Nature*. "Of all the fossil forms known to us, the australopithecinae are the nearest akin to man and the most likely to stand in the direct line of man's ascent," he continued enthusiastically in his book *A New Theory of Human Evolution*, published in 1948. He even went so far as to suggest that the australopithecines be renamed "Dartians" in honor of their original discoverer.

Australopithecus was not renamed, but the scientific turn-around affirming its authenticity and importance in the story of human evolution was nearly unanimous. Australopithecines at last were accepted as important early hominids. And with this shift in opinion came recognition that the development of the brain to its human size had occurred after, and not before, the development of upright walking.

Not *everyone* agreed, though. Even after the disturbing discovery of the Piltdown hoax in the 1950s there were still some evolutionists who continued to hold that the large brain, and not upright walking, had led the way in human evolution. Louis Leakey, one of the most outspoken critics of the place of australopithecines in the human evolutionary line, was even then at work in Africa, attempting to find fossil evidence for his belief that man had been shaped several million years ago and had remained virtually unchanged ever since. *Australopithecus*, Java man, Peking man, Neanderthal, and all the rest, Leakey believed, were just failed evolutionary experiments, at the most cousins or offshoots of humankind that ended in extinction.

In the second half of the 20th century, Leakey, his gifted archaeologist wife Mary, and later their equally famous and brilliant son Richard, would make a series of spectacular finds that, along with equally spectacular australopithecine finds by the young scientist Donald Johanson, would again shake the scientific world with controversy.

Much of the battle would again center around *Australopithecus*. Was it a direct ancestor that eventually became modern man, or just an evolutionary cousin, thriving in parallel with the *Homo* line from some common, yet-to-be-discovered ancestor?

Throughout much of the remainder of the 20th century, arguments would rage as researchers continued their attempt to unravel the evolutionary lineage of the human family.

E P I L O G U E

*T*he year 1945 was a time of great contrasts: relief and optimism now that a devastating war had finally come to an end, set against the stark horror of the bombing of Hiroshima and Nagasaki. For the first time scientists found themselves stepping out from their research institutes and university laboratories to speak out on the political and social responsibilities required of those who possess knowledge. Many physicists, including Niels Bohr and Albert Einstein, became active in causes such as Atoms for Peace, an organization devoted to the development of peaceful uses for atomic energy and the establishment of treaties against the use of atomic weapons. And many biologists found themselves speaking out on the effects of radiation on living things of all kinds, and on humans in particular. The mushroom cloud of the atomic bomb had changed the world—and the world of science—forever.

Science had changed in other ways, as well. In the 50 years since 1895, science had become, more and more, a team effort. Laboratories and institutes had become the nurseries of new developments and discoveries, from the Cavendish Laboratory for physics to the Morgan group's "Fly Room" at Columbia University. Ever-increasing complexities produced new branches of science, such as neuroanatomy and subatomic physics. And scientific progress required more and more expensive and specialized equipment, such as could be financed only through funding from institution grants, corporate research-and-development resources or governments. Gone were the days when Copernicus could sit alone in a tower and calculate the structure of the universe, or when Benjamin Franklin could retire from business and, with no formal training, make major contributions to the understanding of electricity. More than ever before, science required not only the special qualities of objectivity coupled with creativity, but also the ability to exchange ideas, learn from others and profit from training.

In addition, the increased expense of research had a negative effect. Ever more dependent on support from governments and large corporations, science became more frequently tied to special interests and more secretive. The international flow of information, during and after the Manhattan Project, was slowed or blocked completely in some cases, as nations and private interests protected their findings.

But science was, in many ways, more exciting than ever. The first half of the 20th century saw the atom structured and restructured, plumbed to its inner core and split. A new idea called relativity set classical physics—a system of thought that had ruled the physical sciences for two centuries—into a tailspin.

Beyond these changes a whole new universe had opened up in the atomic world: the revolutionary idea of the quantum had changed forever our ideas of an orderly and predictable universe. Matter itself was not a tangible, stable thing but an intricate dance of particles, interactions and, most disturbingly, it seemed, chance.

In the life sciences, powerful new concepts were born: heredity controlled by tiny, microscopic entities within the cells of an organism; "magic bullets" to destroy the enemies of an organism; synthetic replacement of substances usually produced by the body, such as insulin, or by the diet, such as vitamins, to sustain life.

In 1945, much lay ahead. Many questions still remained to be answered—and some important discoveries lay just around the corner. The atomic explorers had only begun to explore the realm of the atom. New instruments and more sophisticated measurements would reveal further extraordinary discoveries beyond even the proton, the neutron and the tiny electron. What new mysteries did the atom have to disclose? What was the connection between atoms and molecules and the basic building blocks of living things? What was the mechanism that enabled chromosomes to replicate themselves? What was the stuff of which genes were made? And how did they carry the blueprint for an offspring's characteristics? How did life originate? How did the universe begin? Is it expanding? Are there other worlds with other living organisms, other civilizations?

The second half of the 20th century in science would reveal an ever more amazing world of unfolding knowledge, a world accessible to its scientists because they were able to stand on the shoulders of the men and women who had lived before them—who loved to ask questions and seek answers.

A P P E N D I X

THE SCIENTIFIC METHOD

*. . . our eyes once opened, . . . we can never go back to the old
outlook. . . . But in each revolution of scientific thought new
words are set to the old music, and that which has gone before
is not destroyed but refocused.*
—A. S. Eddington

What is science? How is it different from other ways of thinking? And
what are scientists like? How do they think and what do they mean
when they talk about "doing science"?

Science isn't just test tubes or strange apparatus. And it's not just frog
dissections or names of plant species. Science is a way of thinking, a vital,
ever-growing way of looking at the world. It is a way of discovering how the

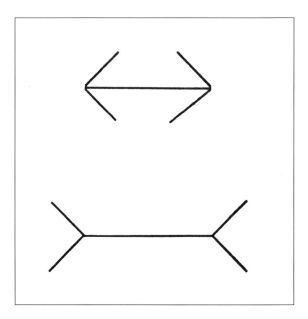

*Looks can be deceiving:
These two lines are the
same length.*

world works—a very particular way that uses a set of rules devised by scientists to help them also discover their own mistakes.

Everyone knows how easy it is to make a mistake about the things you see or hear or perceive in any way. If you don't believe it, look at the two horizontal lines on the previous page. One looks like a two-way arrow; the other has the arrow heads inverted. Which one do you think is longer (not including the "arrow heads")? Now measure them both. Right, they are exactly the same length. Because it's so easy to go wrong in making observations and drawing conclusions, people developed a system, a "scientific method," for asking "How can I be sure?" If you actually took the time to measure the two lines in our example, instead of just taking our word that both lines are the same length, then you were thinking like a scientist. You were testing your own observation. You were testing the information that we had given you that both lines "are exactly the same length." And, you were employing one of the strongest tools of science to do your test: you were quantifying, or measuring, the lines.

Some 2,400 years ago, the Greek philosopher Aristotle told the world that when two objects of different weights were dropped from a height the heaviest would hit the ground first. It was a common-sense argument. After all, anyone who wanted to try a test could make an "observation" and see that if you dropped a leaf and a stone together the stone would land first. Try it yourself with a sheet of notebook paper and a paperweight in your living room. Not many Greeks tried such a test, though. Why bother when the answer was already known? And, being philosophers who believed in the power of the human mind to simply "reason" such things out without having to resort to "tests," they considered such an activity to be intellectually and socially unacceptable.

Centuries later, Galileo Galilei, a brilliant Italian who liked to figure things out for himself, did run some tests. Galileo, like today's scientists, wasn't content merely to observe the objects falling. Using two balls of different weights, a time-keeping device, and an inclined plane, or ramp, he allowed the balls to roll down the ramp and carefully *measured* the time it took. And, he did this not once, but many times, inclining planes at many different angles. His results, which still offend the common sense of many people today, demonstrated that, if you discount air resistance, all objects would hit the ground at the same time. In a perfect vacuum (which couldn't be created in Galileo's time), all objects released at the same time from the same height would fall at the same rate! You can run a rough test of this yourself (although it is by no means a really accurate experiment) by crumpling the notebook paper into a ball and then dropping it at the same time as the paperweight.

Galileo's experiments (which he carefully recorded step by step) and his conclusions based on these experiments demonstrate another important attribute of science. Anyone who wanted to could duplicate the experiments and either verify his results or, by looking for flaws or errors in the experiments, prove him partially or wholly incorrect. No one ever proved Galileo wrong. And years later when it was possible to create a vacuum (even though his experiments had been accurate enough to win everybody over long before that), his conclusions passed the test.

Galileo had not only shown that Aristotle had been wrong. He demonstrated how, by observation, experiment and quantification, Aristotle, if he had so wished, might have proven himself wrong—and thus changed his own opinion! Above all else the scientific way of thinking is a way to keep yourself from fooling yourself—or, from letting nature (or others) fool you.

Of course science is more than observation, experimentation and presentation of results. No one today can read a newspaper or a magazine without becoming quickly aware of the fact that science is always bubbling with *theories*. "ASTRONOMER AT X OBSERVATORY HAS FOUND STARTLING NEW EVIDENCE THAT THROWS INTO QUESTION EINSTEIN'S THEORY OF RELATIVITY," says a magazine. "SCHOOL SYSTEM IN THE STATE OF Y CONDEMNS BOOKS THAT UNQUESTIONINGLY ACCEPT DARWIN'S THEORY OF EVOLUTION," proclaims a newspaper. "BIZARRE NEW RESULTS IN QUANTUM THEORY SAY THAT YOU MAY NOT EXIST!" shouts another paper. What is this thing called *theory?*

Few scientists pretend any more that they make use of the completely "detached" and objective "scientific method" proposed by the philosopher Francis Bacon and others at the dawn of the scientific revolution in the 17th century. This "method," in its simplest form, proposed that in attempting to answer the questions put forward by nature, the investigator into nature's secrets must objectively and without preformed opinions observe, experiment and gather data about the phenomena. After Isaac Newton demonstrated the universal law of gravity, some curious thinkers suggested that he might have an idea *what gravity was.* But he did not see such speculation as part of his role as a scientist. "I make no hypotheses," he asserted firmly. Historians have noted that Newton apparently did have a couple of ideas, or "hypotheses," as to the possible nature of gravity, but for the most part he kept these private. As far as Newton was concerned there had already been enough hypothesizing and too little attention paid to the careful gathering of testable facts and figures.

Today, though, we know that scientists don't always follow along the simple and neat pathways laid out by the trail guide called the "scientific method." Sometimes, either before or after experiments, a scientist will get an idea or a hunch (that is, a somewhat less than well-thought-out hypothesis) that suggests a new approach or a different way of looking at a problem. Then he or she will run experiments and gather data to attempt to prove or disprove this hypothesis. Sometimes the word *hypothesis* is used more loosely in everyday conversation, but for a hypothesis to be valid scientifically it must have built within it some way that it can be proven wrong, if, in fact, it is wrong. That is, it must be falsifiable.

Not all scientists actually run experiments themselves. Most theoreticians, for instance, map out their arguments mathematically. But hypotheses, to be taken seriously by the scientific community, must always carry with them the seeds of falsifiability by experiment and observation.

To become a theory, a hypothesis has to pass several tests. It has to hold up under experiments, not only to the scientist conducting the experiments or making the observations, but to others performing other experiments and observations as well. Then when thoroughly reinforced by continual testing and appraising, the hypothesis may become known to the scientific or popular world as a *theory*.

It's important to remember that even a theory is also subject to falsification or correction. A good theory, for instance, will make "predictions"— events that its testers can look for as a further test of its validity. By the time most well-known theories such as Einstein's theory of relativity or Darwin's theory of natural selection reach the textbook stage, they have survived the gamut of verification to the extent that they have become productive working tools for other scientists. But in science, no theory can be accepted as completely "proven"; it must remain always open to further tests and scrutiny as new facts or observations emerge. It is this insistently self-correcting nature of science that makes it both the most demanding and the most productive of humankind's attempts to understand the workings of nature. This kind of critical thinking is the key element of doing science.

The cartoon-version scientist portrayed as a bespectacled, rigid man in a white coat, certain of his own infallibility, couldn't be farther from reality. Scientists, both men and women, are as human as the rest of us—and they come in all races, sizes and appearances, with and without eyeglasses. As a group, because their methodology focuses so specifically on fallibility and critical thinking, they are probably even more aware than the rest of us of how easy it is to be wrong. But they like being right whenever possible and they like working toward finding the right answers to questions. In most cases that's why they became scientists.

CHRONOLOGY

THE SECOND HALF OF THE NINETEENTH CENTURY

1857   Gregor Mendel begins his eight-year statistical study of peas

1860s   Beginning of quantitative spectral analysis

1882   Walther Flemming publishes his findings about a process he calls "mitosis," the replication of what later come to be called "chromosomes"

1883   Camillo Golgi discovers "Golgi cells," a type of cell in the nervous system

1885   Johann Balmer publishes his formula for frequencies of the hydrogen atom's spectral lines

1887   Heinrich Hertz discovers photoelectric effect

1887   Michelson-Morley experiment to detect ether

1887   Edouard van Beneden demonstrates that the number of chromosomes is constant in every cell in an organism, and notes that each species seems to have a characteristic number of chromosomes per cell

1888   Hertz detects radio waves

1891   George Johnstone Stoney suggests that the fundamental particle of electricity be called an electron

1893   Santiago Ramón y Cajal proposes that learning results from increased connections between neurons

1893   Henry Ford builds his first experimental "horseless carriage"

1894   J. J. Thomson announces that the velocity of cathode rays is much slower than that of light

1894   The discovery of Java man is announced by Eugène Dubois

1895   Sir William Ramsay discovers the element helium on Earth and finds that it would lie between hydrogen and lithium in the periodic table

1895   E. E. Barnard photographs the Milky Way

1895   Hendrik Lorentz independently develops what we now call the Lorentz-FitzGerald contraction

1895   Wilhelm Konrad Röntgen discovers X rays

1895	Jean-Baptiste Perrin demonstrates that cathode rays are charged particles
1895	Lorentz mass-velocity relationship
1895	Cloud chamber developed
1895	First doctoral degree awarded by a German university to a woman on the basis of the regular examination process
1896	Henri Becquerel discovers radioactivity when he observes a strange, penetrating radiation glowing from a sample of uranium salt
1896	Lorentz and Pieter Zeeman discover the Zeeman effect
1897	Edouard Buchner discovers that a cell-free extract of yeast can convert sugar into alcohol (fermentation without the presence of a living cell), marking the beginning of biochemistry. Previously, chemists believed that the processes associated with life could take place only inside living cells
1897	Thomson discovers the electron
1897	Marie Curie demonstrates uranium radiation from uranium atom
1897	Ernest Rutherford names alpha and beta rays
1898	Spanish-American War breaks out in February; treaty signed in December
1898	Martinus Willem Beijerinck, of the Netherlands, announces that tobacco mosaic disease is caused by an infective agent that he names a filterable virus; he becomes the first to identify a virus
1898	Discovery of radium by Marie and Pierre Curie. Marie Curie coins the term *radioactivity* and also detects polonium
1898	Thomson proposes raisins-in-poundcake model of atom
1899	Ernst Heinrich Haeckel publishes the view that the mind, though the result of creation, depends on the body and does not live beyond the death of the body
1899	German-American physiologist Jacques Loeb discovers parthenogenesis when he succeeds in raising unfertilized sea urchin eggs to maturity after changing their environment

THE EARLY TWENTIETH CENTURY

1900	Max Planck's discovery of his law of blackbody radiation marks the beginning of quantum theory
1900	Increase of mass by velocity, which had been predicted by Lorentz (Lorentz-FitzGerald contraction), measured for the first time, but no one has a comprehensive physical theory that explains it

1900   Hugo de Vries discovers Mendel's work on mutations in pea plants and publishes his own findings on mutations; two other scientists independently make similar discoveries and also find Mendel's work the same year

1900   What later will be called gamma rays discovered by Paul Villard

1901   Jokichi Takamine, of Japan, discovers adrenaline and synthesizes it; Thomas Bell independently makes the same discovery

1901   Pierre Curie works on radioactive energy and gets first indication that a new energy source might be hidden somewhere inside the atom

1902   More detailed studies of photoelectric effect by Philipp von Lenard, first noted in 1887

1902   William Bayliss and Ernest Starling establish the importance of hormones when they discover secretin, a hormone that controls the pancreas and is secreted by the walls of the small intestine

1902   Walter S. Sutton points out similarities between the behavior of chromosomes and Mendel's inheritance factors

1903   Orville and Wilbur Wright conduct the first powered flight by a human, near Kitty Hawk, North Carolina

1903   Publication of Sutton's *Chromosomes Theory of Heredity* supports his previous proposal and argues that each egg or sperm cell contains only one of each chromosome pair

1904   Ramón y Cajal establishes the theory that the nervous system is composed only of nerve cells and their processes

1905   Albert Einstein introduces the light-quantum (later called the photon) and develops the special theory of relativity

1906   Frederick Gowland Hopkins publishes the first work on "accessory food factors," which later come to be known as vitamins

1906   San Francisco earthquake and fire, April 18–19

1907   Thomas Hunt Morgan begins his work on fruit flies (*Drosophila melanogaster*) that will prove the role of chromosomes in heredity, establish mutation theory and lead to fundamental understanding of the mechanisms of heredity

1908–09   Perrin works out approximate size of atoms from actual observations—a final demonstration of their existence

1908–09   Rutherford begins (with Hans Geiger) his investigations into the scattering of alpha particles by gold foil

1909   Wilhelm Johannsen, of Denmark, coins the word *gene* to describe the carrier of heredity. He also establishes the words *genotype* (the genetic constitution of an organism) and *phenotype* (the actual organism resulting from the genotype in combination with environmental factors)

1910   Morgan discovers that certain inherited characteristics are sex-linked

1910   After extensive testing, Paul Ehrlich and his associates discover salvarsan (arsphenamine, or "number 606"), a remedy for syphilis and several other diseases

1911   Rutherford presents his discovery of the atomic nucleus

1911   Charles Wilson perfects his cloud chamber

1911   Robert Millikan works out electron charge, for which he will later get a Nobel prize

1911   Victor Franz Hess works on cosmic rays

1912   The *Titanic*, a huge passenger ship, sinks in the North Atlantic on its first voyage; 1,513 people die

1913   Niels Bohr publishes his first papers on his model of the atom and presents his theory for the first time to an international audience, in Birmingham, England

1913   Johannes Stark announces the splitting of spectral lines when atomic hydrogen is exposed to a static electric field (the Stark effect)

1913   Henry Moseley's experiments bring definitive order in the periodic table of the elements

1913   Bohr presents a lecture to the Danish Physical Society that contains the germs of the correspondence principle

1914   Assassination of the Austrian archduke, Francis Ferdinand, and his wife at Sarajevo precipitates World War I

1914   Franck-Hertz experiment confirms Bohr's ideas of quantum jumps

1914   Moseley's work on atomic number

1914   Father-and-son team, William Henry Bragg and William Lawrence Bragg, determine wavelength of X rays

1914   Rutherford proposes his proton theory

1915   The Panama Canal is officially opened, July 12

1915   *The Mechanism of Mendelian Heredity*, by Thomas Hunt Morgan, Calvin Bridges, Alfred Sturtevant and Hermann Muller, is published and becomes a classic in genetics

1915–16 Discovery of bacteriophages (bacteria-eating viruses) by Frederick William Twort and, independently, by Félix-Hubert d'Hérelle

1916   Einstein's theory of general relativity

1916   Arnold Sommerfeld elaborates on Bohr atom by working out that electrons must move in elliptical orbits and not circular ones. Sommerfeld's theory of the fine structure of the spectral lines in hydrogen evolves from this

1916–17 Einstein introduces probabilities into quantum dynamics

1916–22 Vain efforts to understand the atomic spectrum of helium
1919 Rutherford observes the first artificial transmutations of elements and finds the first indications that nuclear forces are not purely electromagnetic
1919 Tests for gravitational deflection of light and Einstein's general theory of relativity
1920 Sommerfeld introduces a fourth quantum number
1921 Einstein receives Nobel prize in physics
1922 Bohr receives Nobel prize in physics
1922 Frederick Banting and Charles Best announce their isolation of insulin
1923 Dirk Coster and György Hevesy announce the discovery of the element hafnium
1923 Louis de Broglie associates waves with electrons (wave-particle duality). He publishes this theory as his Ph.D. thesis. Einstein sees it and thinks it very important. Later experimental evidence (see 1927) proves him right. This is one of the major linchpins and quandaries of quantum theory
1924 Hendrik Kramers and John Slater present their theory that in atomic processes energy is only conserved statistically: experiments prove them wrong later in year
1925 Wolfgang Pauli enunciates the exclusion principle
1925 Werner Heisenberg's first paper on quantum mechanics (matrix mechanics)
1925 First comprehensive treatment of matrix mechanics, by Max Born and Werner Heisenberg
1925 John Scopes, a Tennessee teacher, put on trial for teaching evolution
1925 George Uhlenbeck and Samuel Goudsmit announce the first theory of electron spin
1926 Pauli derives the Balmer formula from matrix mechanics
1926 Erwin Schrödinger's first paper on wave mechanics. It is an implicit attack on Heisenberg's matrix mechanics. But what is it that is "waving"? Schrödinger thinks actual small packets of waves
1926 Enrico Fermi's first paper on "Fermi-Dirac statistics"
1926 Heisenberg resolves the mystery of helium spectrum
1926 Max Born introduces for the first time the quantum mechanical probability concept and disagrees with Schrödinger's "waves"
1926 Paul Dirac gives the quantum mechanical foundations of quantum statistics and of Planck's radiation law
1926 Dirac's paper on the transformation theory of quantum mechanics

1927	Dirac's first paper on quantum electrodynamics
1927	Heisenberg's paper on the uncertainty principle
1927	Bohr presents complementarity concept
1927	Beginning of Einstein-Bohr dialog at Solvay
1927	Clinton Davisson and George Thomson (son of J. J. Thomson) independently obtain experimental evidence for de Broglie's wave-particle duality theory
1928	Dirac discovers the relativistic wave equation of the electron
1928	Alexander Fleming accidentally discovers a mold he calls penicillin, which has unusual anti-bacterial qualities
1929	J. D. Cockcroft and E. T. S. Walton devise first particle accelerator, or "atom smasher"
1929	U.S. stock market crashes; Great Depression begins
1930	Sixth Solvay Conference. Einstein offers his "clock in the box" thought experiment. Bohr refutes it. Einstein has begun his battle with the idea that quantum mechanics is a complete theory. He does not like the probability aspects—will come to insist that "God does not play dice with the Universe"
1930	Dirac proposes existence of antimatter
1931	Pauli proposes the existence of the neutrino
1931	First report of functioning cyclotron in Berkeley, California
1931	Dirac postulates the positron
1931	George Gamow completes first textbook on theoretical nuclear physics
1931	Robert van de Graaff's electrostatic generator
1931	First experimental evidence for the positron
1931	Fermi names "neutrino" (Pauli had predicted its existence), though it is not detected until 1956
1932	Discovery of the neutron by James Chadwick
1932	First nuclear process produced by an accelerator by Cockroft and Walton
1932	Carl David Anderson studies cosmic rays
1932	Anderson discovers positron (which Dirac had predicted)
1932	Gerhard Domagk develops Prontosil, the first of the "sulfa" (sulfonamide) drugs
1933	German government strips Jews of official positions
1934	Discovery of artificial radioactivity by Irène and Frédéric Joliot-Curie
1934	Rutherford and associates bombard deuterium with deuterons and produce tritium—producing the first nuclear fusion reaction
1936	Bohr's theory of compound atomic nucleus

1938   Ernst B. Chain and Howard Walter Florey isolate and purify penicillin for general clinical use from the mold discovered ten years earlier by Fleming. With Fleming, they are awarded the 1945 Nobel prize for physiology or medicine for their work

1938   Otto Hahn and Fritz Strassmann discover that irradiation of uranium with neutrons yields barium. Lise Meitner and Otto Frisch interpret this result as uranium fission. Bohr and Fermi's reports on fission at a physics meeting in Washington, D.C. mark the beginning of widespread activities in fission research

1939   World War II breaks out

1941   U.S. president Franklin D. Roosevelt decides to go ahead with atomic weapons research

1941   U.S. military base at Pearl Harbor, Hawaii bombed by Japanese; United States enters war

1942   The Manhattan Project begun in United States. J. Robert Oppenheimer at helm in successful attempt to build the world's first atomic bomb

1942   Fermi in Chicago builds first nuclear reactor and gets a chain reaction, thereby beginning the atomic age

1945   August 6, atomic bomb dropped by United States on Hiroshima; August 9, atomic bomb dropped by United States on Nagasaki

1945   September 2, World War II ends

G L O S S A R Y

absolute magnitude   a measure of the brightness that a star would have at a standard distance of 10 parsecs (a little over 190 trillion miles). Magnitude is a measure of brightness of stars and other celestial objects

alpha particle   helium nucleus; a positively charged particle consisting of two protons and two neutrons emitted by radioactive elements during the process of radioactive decay

apparent magnitude   a measure of the brightness of light from a star or other object in the sky as measured from Earth. However, because a very bright but distant star could have the same apparent magnitude as a dimmer, closer star, the intrinsic brightness, or luminosity, of a star is measured in terms of *absolute magnitude*

atom   the smallest chemical unit of an element, consisting of a dense, positively charged nucleus surrounded by negatively charged electrons. The Greek thinker Leucippus and his student Democritus originally conceived of the idea of the atom in the 5th century B.C. as the smallest particle into which matter could be divided. (The word *atom* comes from the Greek word *atomos*, which means "indivisible.") But in the 1890s and early 20th century, scientists discovered that the atom is made up of even smaller particles, most of which are very strongly bound together

atomic bomb   an explosive weapon having enormous destructive power, deriving its force from the rapid release of energy from the fission or splitting of the atomic nucleus of heavy elements, such as uranium and plutonium

atomic mass   (also known as atomic weight) the relative average mass in atomic mass units (or amu, equal to 1/12 the weight of the isotope carbon-12) of the masses of all the isotopes found in a natural sample of an element

atomic spectrum   the characteristic pattern produced by light emitted by any element. For example, when electrical energy is passed through an element enclosed in a gas tube, light is given off. If you pass this light through a prism, the light separates into a series of specific colors with spaces between them. This pattern is that element's atomic spectrum, or emission line spectrum. These spectra can be used as a sort of fingerprint of the elements. Major color or emission lines for some elements are striking and easily detected; for example, the main emission line for sodium is yellow; for copper, green; for potassium, violet, and so on

beta particle   a high-speed electron, or positron, emitted during radioactive decay

binding energy   the amount of energy required to hold the parts of an atomic nucleus together

chain reaction   a multistage, self-sustaining nuclear reaction, consisting of a series of fissions. The reaction continues in a chain or series because the average number of neutrons produced exceeds the number absorbed or lost during a given period of time

chromosome   a linear body found in the nucleus of all plant and animal cells that is responsible for determining and transmitting hereditary characteristics

cosmic rays   a stream of ionizing radiation that enters the Earth's atmosphere from outer space. The rays consist of high-energy atomic nuclei, alpha particles, electromagnetic waves and other forms of radiation

critical mass   an exact amount of a radioactive material, such as uranium and plutonium, needed to produce a chain reaction

cyclotron   a device used to accelerate subatomic particles to immense speed along a spiral path by means of a fixed magnetic field and a variable electric field. Cyclotrons have been used, for example, to separate different isotopes of uranium

deuterium   an isotope of hydrogen having an atomic weight of 2.0141, used in the production of heavy water. Its nucleus contains both a proton and a neutron (the more common hydrogen nucleus contains only one proton)

Drosophila   a fruit fly used extensively in genetic studies because it has only four pairs of chromosomes and breeds rapidly

electron   a very light, negatively charged subatomic particle that orbits the nucleus of an atom

electron volt   the amount of energy necessary to accelerate an electron through a potential difference of one volt

fission   a nuclear reaction in which an atomic nucleus splits into fragments. Fission can generate enormous amounts of energy

fusion a reaction in which atomic nuclei combine to form more massive nuclei; the process usually creates some excess mass that is released as energy

gamma ray an electromagnetic radiation that is massless and moves at the speed of light, with energy greater than several hundred thousand electron volts. It is emitted when an excited atomic nucleus passes to a lower energy state

gene a hereditary unit that is located in a fixed place on a chromosome and has a specific influence on the physical makeup of a living organism

half-life the time it takes for one-half the nuclei in a given amount of a radioactive isotope to decay into another substance. That is, after one half-life, one-half of the original sample of a radioactive element will have transformed. Half-lives of different isotopes range from a few millionths of a second to several billion years

hereditary characteristic a trait that can be transmitted by genes from generation to generation

hominid a member of the family of two-legged primates that includes several human species, both extinct and living, among them *Australopithecus*, *Homo erectus* and *Homo sapiens*

hominoid a member of the superfamily that includes both hominids and the great apes

Homo erectus the genus and species names given to Eugène Dubois's Java man, to Peking man and to other extinct, apelike humans. Large (six feet tall), having heavy browridges and a brain measuring one-half to two-thirds the size of the modern human brain, *Homo erectus* lived between .5 million and 1.6 million years ago

Homo sapiens the species that includes both Neanderthals and modern humans

hydrogen bomb a weapon the destructive power of which comes from the sudden release of energy that takes place during the fusion of hydrogen and deuterium nuclei (containing one proton) into helium and tritium nuclei (containing two protons). The hydrogen bomb uses the tremendous burst of heat from an exploding fission bomb to trigger the fusion process

implosion a bursting inward (instead of outward, as in an explosion)

isotope one of two or more of any element's atoms with the same number of protons in the nuclei but different numbers of neutrons, having the same chemistry, but different nuclear physics. Even though these are atoms of the same element, they have different properties because they have different numbers of neutrons in their nuclei. For example U-235, an isotope of uranium, is more fissionable than U-238 because it has three less neutrons in its nucleus

mutation   any alteration in an organism that can be inherited, carried by a gene or group of genes

natural selection, theory of   Charles Darwin's theory, which he called the "struggle for survival," that those organisms less well adapted to reproduce in their environment tend to become extinct, while those that are better adapted to reproduce tend to survive. Natural selection acting on a varied population, according to Darwin, results in evolution

neutron   an uncharged, or neutral, particle found in the nucleus of every atom except hydrogen

photon   a small particle, or "energy packet," of electromagnetic energy, having no mass and no electric charge. The photon's existence was first proposed by Albert Einstein to explain the behavior of light in photoelectric experiments

plutonium   an artificially created radioactive element, which does not appear naturally on Earth. First created by American chemist Glenn Seaborg in 1940. It has an atomic number of 94, and its most common isotope is Pu-244. It is used as a fuel in atomic reactors and in nuclear weapons

positron   a positively charged electron, the mirror image of an electron

proton   a positively charged particle found in the nucleus of all atoms

quantum   a small, indivisible unit, or packet, of energy

quantum theory   a theory of atomic and subatomic interaction based on the behavior of the photon as both a wave and a particle. The theory makes use of the concepts of probability and quantum energy to explain how the atom is held together and how it functions

radioactivity   spontaneous emission of radiation, or rays, either directly from unstable atomic nuclei or as a result of a nuclear reaction. The rays given off can be neutrons, alpha particles, beta particles or gamma rays

subatomic particles   the extremely small particles of which atoms are composed, including neutrons, protons, electrons and positrons, gamma rays and neutrons

uranium   a heavy, silvery white radioactive element used in nuclear fuels, nuclear weapons and research; the heaviest naturally occurring element

X ray   a relatively high-energy photon, or stream of photons, with a very short wavelength. X rays are emitted from a metal target when it is struck by electrons used to bombard a target

F U R T H E R
R E A D I N G

ABOUT SCIENCE:

Cole, K. C. *Sympathetic Vibrations: Reflections on Physics as a Way of Life.* New York: William Morrow and Co., 1985. Well-written, lively and completely intriguing look at physics presented in a thoughtful and insightful way by a writer who cares for her subject. The emphasis here is primarily on modern physics; concentrates more on the ideas than on the history.

Gardner, Martin. *Fads and Fallacies in the Name of Science.* New York: New American Library, 1986 (reprint of 1952 edition). A classic look at pseudoscience by the master debunker. Includes sections on pseudo-scientific beliefs in the 19th century.

Gonick, Larry, and Art Huffman. *The Cartoon Guide to Physics.* New York: Harper Perennial, 1991. Fun, but the whiz-bang approach sometimes zips by important points a little too fast.

Hann, Judith. *How Science Works.* Pleasantville, N.Y.: The Reader's Digest Association, Inc., 1991. Lively, well-illustrated look at physics for young readers. Good, brief explanations of basic laws and short historical overviews accompany many easy experiments readers can perform.

Hazen, Robert M., and James Trefil. *Science Matters: Achieving Scientific Literacy.* New York: Doubleday, 1991. A clear and readable overview of basic principles of science and how they apply to science in today's world.

Holzinger, Philip R. *The House of Science.* New York: John Wiley and Sons, 1990. Lively question-and-answer discussion of science for young adults. Includes activities and experiments.

Trefil, James. *1001 Things Everyone Should Know about Science.* New York: Doubleday, 1992. The subtitle, *Achieving Scientific Literacy*, tells all. Well done for the average reader but includes little history.

ABOUT THE PHYSICAL SCIENCES, 1895–1945:

Bergmann, Peter G. *The Riddle of Gravitation*. New York: Charles Scribner's Sons, 1968. A classic explanation of Einstein's theories. Easygoing and easy to read.

Boorse, Henry A., Lloyd Motz, and Jefferson Hane Weaver. *The Atomic Scientists: A Biographical History*. New York: John Wiley and Sons, Inc., 1989. Interesting approach to the history of atomic science. The development is followed through a series of brief but lucid biographies of the major figures. May be a little difficult without some background.

Calder, Nigel. *Einstein's Universe*. New York: Crown Publishers, 1979. An easy-to-read look at both the special and general theories of relativity.

Cassidy, David C. *Uncertainty: The Life and Times of Werner Heisenberg*. New York: W. H. Freeman and Co., 1992. Sometimes heavy-going, but authoritative biography of Heisenberg.

Clark, Ronald W. *Einstein: The Life and Times*. New York: World Publishing, 1971. Standard biography, well done, but a little on the long and dry side.

Cline, Barbara Lovett. *Men Who Made the New Physics*. Chicago: University of Chicago Press, 1987. Very easy-to-read narrative study of the major figures in the quantum revolution in physics.

Crease, Robert P., and Charles C. Mann. *The Second Creation: Makers of Revolution in 20th-Century Physics*. New York: Macmillan Publishing Co., 1986. Thorough, well-done study, but sometimes a little heavy going for the average reader.

d'Abro, A. *The Rise of the New Physics: In Two Volumes*. New York: Dover, 1951. A standard, but sometimes dry, history. Contains good background information on the early pioneers and theories, though sometimes a little dated.

Einstein, Albert. *Out of My Later Years*. New York: Philosophical Library, 1950. Essays on science, humanity, politics and life by the gentle-thinking Einstein.

Fermi, Laura. *Atoms in the Family*. Chicago: University of Chicago Press, 1954. Informative and easy-to-read remembrances by the talented wife of Enrico Fermi.

French, A. P. *Einstein: A Centenary Volume*. Cambridge, Mass.: Harvard University Press, 1979. Major scientists, thinkers and friends remember the great man in a variety of ways. Some essays easy to follow, some much more esoteric and difficult. All in all a good look at Einstein and his major contributions.

French, A. P., and P. J. Kennedy, eds. *Niels Bohr: A Centenary Volume*. Cambridge, Mass.: Harvard University Press, 1985. A well-rounded look at Niels Bohr, told by friends, students and coworkers. Lively and easy to read.

Frisch, Otto. *What Little I Remember.* Cambridge, England: Cambridge University Press, 1979. Casual, highly readable reminiscences by a participant in the early days of atomic physics.

Gamow, George. *The Great Physicists from Galileo to Einstein.* New York: Dover Publications Inc., 1988. Gamow takes a look at some major historical physicists and their work. As always with Gamow, readable and enlightening. A little dated but the essential information remains valuable.

———. *Thirty Years That Shook Physics.* New York: Doubleday and Co., Inc., 1966. Well-done look at the history and development of quantum physics. A little dated but still relevant and informative about the early days of quantum theory.

Han, M. Y. *The Secret Life of the Quanta.* Blue Ridge Summit, Penn.: Tab Books, 1990. Coming at the quanta from the practical rather than the theoretical end, Han concentrates on explaining how our knowledge of the quanta allows us to build such high-tech devices as computers, lasers and the CAT Scan—while not neglecting theory and the bizarre aspects of his subject.

———. *The Probable Universe.* Blue Ridge Summit, Penn.: Tab Books, 1993. Han follows up his successful *Secret Life of the Quanta* with a closer look at the mysteries and controversies, as well as the theoretical and technological successes, of quantum physics.

Hazen, Robert M., and James Trefil. *Science Matters: Achieving Scientific Literacy.* New York: Doubleday, 1991. Includes good brief discussions on relativity and quantum physics.

Hendry, John. *The Creation of Quantum Mechanics and the Bohr-Pauli Dialogue.* Dordrecht, The Netherlands: D. Reidel Publishing Company, 1984. Tough going but contains a wealth of information for the well-informed reader.

Hoffmann, Banish. *Albert Einstein: Creator and Rebel.* New York: Viking Press, 1972. Wonderful, warm and easy-to-read look at Einstein by a very readable writer.

———. *The Strange Story of the Quantum.* New York: Dover Publications, 1959. Accessible and intriguing as always with Hoffmann's work.

Holton, Gerald. *Thematic Origins of Scientific Thought.* Revised Edition. Cambridge, Mass.: Harvard University Press, 1988. A good, standard study but may be tough reading for the average reader.

Jones, Roger S. *Physics for the Rest of Us.* Chicago: Contemporary Books, 1992. Informative, non-mathematical discussions of relativity and quantum theory told in a straightforward manner, along with discussions on how both affect our lives and philosophies.

Kunetka, James W. *City of Fire: Los Alamos and the Birth of the Atomic Age, 1943–1945.* Englewood Cliffs, N.J.: Prentice-Hall, 1978. A strong, readable narrative—recommended reading.

Lamer, Max. *Conceptual Development of Quantum Mechanics.* New York: McGraw-Hill, 1966. Superior study but tough going without some familiarity with the subject.

Lederman, Leon M., and David N. Schramm. *From Quarks to the Cosmos: Tools of Discovery.* New York: Scientific American Library, 1989. Takes a good look not only at the theorists but at the experimenters and experiments in the development of our understanding of the atom. Well-done and easy-to-follow book.

Lederman, Leon, with Dick Teresi. *The God Particle.* New York: Houghton Mifflin, 1993. A sprightly and often humorously personal look at the history of physics and particle physics. Easy to read, lucid and informative.

Lightman, Alan. *Einstein's Dreams.* New York: Pantheon Books, 1993. Wonderful, readable look at time as seen by Einstein, set as a series of vignettes, each taking a different look at the mysteries of time. A completely charming and thoughtful book.

McCormmach, Russell. *Night Thoughts of a Classical Physicist.* Cambridge, Mass.: Harvard University Press, 1982. A fascinating novel treating the mind and emotions of a turn-of-the-century physicist in Germany who attempts to understand the disturbing changes happening in physics and the world around him.

Miller, Arthur I. *Imagery in Scientific Thought.* Cambridge, Mass.: The MIT Press, 1986. Historical and philosophical look at the development in the thinking of physics. Might be a little abstract for the average reader.

Moore, Ruth. *Niels Bohr: The Man, His Science, and the World They Changed.* New York: Alfred A. Knopf, 1966. A readable biography of Bohr. Dependable but a little dated.

Moore, Walter. *Schrödinger: Life and Thought.* Cambridge, England: Cambridge University Press, 1992. A look at the unorthodox life of one of the major figures in the quantum revolution.

Morris, Richard. *The Nature of Reality.* New York: McGraw-Hill, 1987. A tough-minded, critical and easy-to-read look at the direction the science of physics is going today.

Pagels, Heinz R. *The Cosmic Code: Quantum Physics as the Language of Nature.* New York: Simon and Schuster, 1982. Quickly becoming a standard reference for the history and problems of quantum physics. Brilliantly written and easy to read, offering a clear look at some of the quantum paradoxes.

Pais, Abraham. *Inward Bound: Of Matter and Forces in the Physical World.* Oxford: Clarendon Press, 1986. Very well-done study but not always an easy read.

———. *Niels Bohr's Times: In Physics, Philosophy and Polity.* Oxford: Clarendon Press, 1991. In-depth, very detailed, and well-organized biography of the great physicist. The attention to detail, though, requires some background.

———. *Subtle Is the Lord: The Science and Life of Albert Einstein.* New York: Oxford University Press, 1982. Pais does his usual intensive job, again best for the well-informed reader, but highly recommended.

Peat, David F. *Einstein's Moon: Bell's Theorem and the Curious Quest for Quantum Reality.* Chicago: Contemporary Books, Inc., 1990. A popularly written study of some of the major paradoxes of quantum theory.

Pelaum, Rosalind. *Grand Obsession: Madame Curie and Her World.* New York: Doubleday, 1989. A fine account of Marie Curie's extraordinary life and work in nuclear physics, as well as some coverage of the life and work of her daughter and son-in-law, Irène and Frédéric Joliot-Curie. Many photographs.

Rhodes, Richard. *The Making of the Atomic Bomb.* New York: Simon and Schuster, 1986. A truly major work covering not only the development of the bomb but the important work in physics leading up to it. Very well done, not too difficult to read, and always fascinating and informative.

Rowland, John. *Rutherford: Atom Pioneer.* New York: Philosophical Library, 1957. A readable biography.

Rozental, S., ed. *Niels Bohr: His Life and Work as Seen by His Friends and Colleagues.* New York: John Wiley and Sons, 1967. Warm, intelligent, and informative recollections.

Segrè, Emilio. *From X-Rays to Quarks: Modern Physicists and Their Discoveries.* San Francisco: W. H. Freeman and Co., 1980. Good study, sometimes tough going.

Swenson, Lloyd S. *The Ethereal Aether.* Austin: University of Texas Press, 1972. A look at the history of the search for the mysterious ether.

Von Baeyer, Hans Christian. *Taming the Atom.* New York: Random House, 1992. Well-told, easy-to-read and up-to-date look at the history and development of atomic physics. Von Baeyer also takes a look at the most recent developments in the field and present and upcoming attempts at experiments to help clear up the quantum mysteries.

Weisskopf, Victor F. *The Joy of Insight: Passions of a Physicist.* New York: Basic Books, 1991. Delightful and personal reminiscences by an easygoing and thoughtful physicist who knew many of the great quantum pioneers first-hand.

———. *The Privilege of Being a Physicist.* New York: W. H. Freeman and Co., 1989. Reminiscences and random thoughts by a well-respected physicist.

ABOUT THE LIFE SCIENCES, 1895–1945:

Asimov, Isaac. *A Short History of Biology.* Garden City, N.Y.: The Natural History Press, 1964. A brief overview of discoveries in biology from earliest times to the 1960s, written clearly, in a blow-by-blow style.

Chase, Allan. *Magic Shots.* New York: William Morrow and Company, Inc., 1982. The story of the quest for vaccinations to prevent infectious diseases in the 19th and 20th centuries, from Edward Jenner's smallpox vaccination to polio vaccines and the ongoing search for cancer vaccines.

Darwin, Charles. *On the Origin of Species by Means of Natural Selection, or the Preservation of Favoured Races in the Struggle for Life.* London: Murray, 1859. One of the most influential books ever published in biology. Highly influential due to the great array of evolutionary evidence Darwin presented. The book also stimulated constructive thinking on a subject that had been notoriously vague and confusing prior to Darwin's time.

Facklam, Margery, and Howard Facklam. *Healing Drugs: The History of Pharmacology.* New York: Facts On File, 1992. Part of the Facts On File Science Sourcebook series, this book tells the stories of Paul Ehrlich and his magic bullet, Best and Banting and their research on insulin, and Fleming's discovery of penicillin, as well as more recent drug research.

Kornberg, Arthur. *For the Love of Enzymes: The Odyssey of a Biochemist.* Cambridge, Mass.: Harvard University Press, 1989. One of the premier biologists of our time provides the 19th- and early 20th-century background for recent advances in biology, as background to his autobiographical account of his own significant work.

Lambert, David, and the Diagram Group. *The Field Guide to Early Man.* New York: Facts On File, 1987. Well-designed and carefully illustrated "handbook" on all aspects of early humans. A wonderful book for finding facts, or just reading for fun. Easy to read and endlessly fascinating.

Lewin, Roger. *Bones of Contention: Controversies in the Search for Human Origins.* New York: Simon and Schuster, 1987. Although primarily concerned with the controversies and developments in the latter half of the 20th century, this book does have good, readable background material about the search for human origins during the period from 1895 to 1945, as well.

Milner, Richard. *The Encyclopedia of Evolution.* New York: Facts On File, 1990. Just what the title says, written in an easygoing, informal style that makes it a joy to use for research or just endless hours of browsing fun. Plenty of surprises, too, including looks at popular films dealing with the subject and lots of little-known oddities.

Moore, John A. *Science As a Way of Knowing: The Foundations of Modern Biology.* Cambridge, Mass.: Harvard University Press, 1993. A cultural history of biology as well as a delightful introduction to the procedures and values of science, written by a leading evolutionary-developmental biologist. Nontechnical and readable, it is excellent reading for the nonspecialist who wants to know more about how modern molecular biology, ecology and biotechnology came to be.

Moore, Ruth. *The Coil of Life: The Store of the Great Discoveries in the Life Sciences.* New York: Alfred A. Knopf, 1961. Engaging account of the events leading up to and beyond the discovery of the structure of DNA in the 1950s.

Morgan, Thomas H., A. H. Sturtevant, C. B. Bridges, and H. J. Miller. *The Mechanism of Mendelian Heredity.* 1915. Reprinted with a new introduction by Garland E. Allen. New York: Johnson Reprint, 1972. An analysis and synthesis of Mendelian inheritance as formulated from the epoch-making investigations of the authors. A classic work that stands as a cornerstone of our modern interpretation of heredity.

Portugal, Franklin H., and Jack S. Cohen. *A Century of DNA.* Cambridge, Mass.: The MIT Press, 1977. Provides a thorough look at the background leading up to the discovery of the role played by DNA in heredity.

Reader, John. *The Rise of Life.* New York: Alfred A. Knopf, 1986. Beautifully illustrated popular-level book covering the origins and evolution not only of humans but of all life forms. Very well done.

Shapiro, Gilbert. *A Skeleton in the Darkroom: Stories of Serendipity in Science.* San Francisco: Harper & Row, 1986. Engagingly told stories of scientific discovery, including a fascinatingly detailed account of Alexander Fleming's initial discovery of penicillin.

Shorter, Edward, Ph.D. *The Health Century: A Companion to the PBS Television Series.* New York: Doubleday, 1987. An easygoing account of the scientific pursuit of keys to better health in the 20th century, set in the social context of the times.

Spencer, Frank. *Piltdown: A Scientific Forgery.* London: Natural History Museum Publications, Oxford University Press, 1990. Exhaustive but highly readable look at the Piltdown hoax and just about everyone involved in the story, including the most likely suspects. Illustrations and seldom-seen photographs help bring the story to life. A must for anyone interested in science and the Piltdown mystery.

Trinkaus, Erik, and Shipman, Pat. *The Neandertals.* New York: Alfred A. Knopf, 1992. Up-to-date information on the once poorly understood Neanderthals. A good, solid book, nicely researched, but a little difficult to read.

I N D E X

Boldface type indicates major topic.
Italic type indicates illustrations.